Africa's Green Revolution

This volume examines the dominant neoliberal agenda for agricultural development and hunger alleviation in Africa. The text reviews the history of African agricultural and food security policy in the post-colonial period, across a range of geographical contexts, in order to contextualise the productionist approach embedded in the much heralded New Green Revolution for Africa. This strategy, supported by a range of international agencies, promotes the use of hybrid seeds, fertilisers, and pesticides to boost crop production. The approach is underpinned by a new and unprecedented level of public–private partnerships as donors actively work to promote the private sector and build links between African farmers, input suppliers, agro-dealers, agro-processors, and retailers. On the consumer end, increased supermarket penetration into poorer neighbourhoods is proffered as a solution to urban food insecurity. The chapters in this volume complicate understandings of this new approach and raise serious questions about its effectiveness as a strategy for increasing food production and alleviating poverty across the continent.

This book is based on a special issue of the *African Geographical Review*.

William G. Moseley is a professor of geography and African studies at Macalester College in Saint Paul, MN, USA. His research is focused on food security, agriculture, development, and environment in the African context.

Matthew A. Schnurr is an associate professor in the Department of International Development Studies at Dalhousie University in Nova Scotia, Canada. He is an environmental geographer with research and teaching interests in environment and development, political ecology, agricultural biotechnology, farmer decision-making, and environmental justice.

Rachel Bezner Kerr is an associate professor in the Department of Development Sociology at Cornell University in Ithaca, NY, USA. Her research interests focus on farmer-led agroecology, critical examinations of neoliberal approaches to agriculture, and the historical, political, economic, and gender dynamics that shape food security in sub-Saharan Africa.

Africa's Green Revolution

Critical perspectives on new agricultural technologies and systems

Edited by
William G. Moseley, Matthew A. Schnurr and Rachel Bezner Kerr

Routledge
Taylor & Francis Group
LONDON AND NEW YORK

First published 2016
by Routledge
2 Park Square, Milton Park, Abingdon, Oxon, OX14 4RN, UK

and by Routledge
711 Third Avenue, New York, NY 10017, USA

Routledge is an imprint of the Taylor & Francis Group, an informa business

British Library Cataloguing in Publication Data
A catalogue record for this book is available from the British Library

ISBN 13: 978-1-138-18595-1

Typeset in Times New Roman
by RefineCatch Limited, Bungay, Suffolk

Publisher's Note
The publisher accepts responsibility for any inconsistencies that may have arisen during the conversion of this book from journal articles to book chapters, namely the possible inclusion of journal terminology.

Disclaimer
Every effort has been made to contact copyright holders for their permission to reprint material in this book. The publishers would be grateful to hear from any copyright holder who is not here acknowledged and will undertake to rectify any errors or omissions in future editions of this book.

Contents

Citation Information

The following chapters were originally published in the *African Geographical Review*, volume 34, issue 1 (April 2015). When citing this material, please use the original page numbering for each article, as follows:

Chapter 1

Introduction: Interrogating the technocratic (neoliberal) agenda for agricultural development and hunger alleviation in Africa
William G. Moseley, Matthew A. Schnurr and Rachel Bezner Kerr
African Geographical Review, volume 34, issue 1 (April 2015) pp. 1–7

Chapter 2

The social realities of technology transfer: smallholder farmers' encounter with a new rice variety
Daniel Bornstein
African Geographical Review, volume 34, issue 1 (April 2015) pp. 8–12

Chapter 3

A political ecology of high-input agriculture in northern Ghana
Hanson Nyantakyi-Frimpong and Rachel Bezner Kerr
African Geographical Review, volume 34, issue 1 (April 2015) pp. 13–35

Chapter 4

Implications of supermarket expansion on urban food security in Cape Town, South Africa
Stephen Peyton, William G. Moseley and Jane Battersby
African Geographical Review, volume 34, issue 1 (April 2015) pp. 36–54

Chapter 5

Should I stay or should I go? Incorporating a commitment to fieldwork throughout an academic career
Kristal Jones, Matthew A. Schnurr, Edward R. Carr and William G. Moseley
African Geographical Review, volume 34, issue 1 (April 2015) pp. 55–68

Chapter 6

Food and nutrition assistance to HIV-infected and affected populations in Ghana: a situational analysis and stakeholder views
Amos Laar, Angela El-Adas, Richard N. Amenyah, Kyeremeh Atuahene, Elizabeth Asare, Eric Y. Tenkorang, Matilda Laar, Andrew Anthony Adjei and Isabella Quakyi
African Geographical Review, volume 34, issue 1 (April 2015) pp. 69–82

Chapter 7

A field assessment of land use systems and soil properties at varied landscape positions in a fragile ecosystem of Mount Elgon, Uganda
Tonny J. Oyana, Ellen Kayendeke, Yazidhi Bamutaze and Danielson Kisanga
African Geographical Review, volume 34, issue 1 (April 2015) pp. 83–103

The following chapter was originally published in *Agroecology and Sustainable Food Systems*, volume 38, issue 6 (November 2014). When citing this material, please use the original page numbering for each article, as follows:

Chapter 8

Smallholder Perspectives on Soil Fertility Management and Markets in the African Green Revolution
Imogen Ruby Vanessa Bellwood-Howard
Agroecology and Sustainable Food Systems, volume 38, issue 6 (November 2014) pp. 660–685

INTRODUCTION

Interrogating the technocratic (neoliberal) agenda for agricultural development and hunger alleviation in Africa

William G. Moseley[a], Matthew A. Schnurr[b] and Rachel Bezner Kerr[c]

[a]*Geography, Macalester College, MN, USA;* [b]*International Development Studies, Dalhousie University, Halifax, Nova Scotia, Canada;* [c]*Development Sociology, Cornell University, Ithica, NY, USA*

This paper introduces a special issue that critically examines the dominant technocratic, neoliberal agenda for agricultural development and hunger alleviation in Africa. We briefly review the history of African agricultural and food security policy in the post-colonial period in order to contextualise the productionist approach embedded in the New Green Revolution for Africa, a strategy comprising the use of hybrid seeds, fertilisers, and pesticides to boost crop production. This approach is underpinned by a new and unprecedented level of public–private partnerships as donors actively work to promote the private sector and build links between African farmers, input suppliers, agro-dealers, agro-processors, and retailers. On the consumer end, increased supermarket penetration into poorer neighbourhoods is proffered as a solution to urban food insecurity. The papers in the special issue complicate understandings of this new approach and raise serious questions about its effectiveness as a strategy for increasing food production and alleviating hunger across the continent.

Introduction

While languishing on the back burner of African development agendas for years, agriculture and food security are now leading priorities with significant donor interest (Bourne, 2014). The twin goals of increased food production and a better fed African population are being conceptualised in a manner, and supported by a constellation of actors and donors, that is simultaneously new and old. This new mix of ideas and players is reorganising African development as we know it. Indeed, actors in this arena often refer to Africa as the last frontier of agricultural development. It is simultaneously a region of vast untapped agricultural potential, a zone of capitalist agricultural expansion, and the final pocket of global hunger to be conquered. The synergies of capitalism and philanthropy, globalisation and humanitarianism, are inseparable. Scholars, of course, would be remiss not to study this unfolding melee of activity. We believe geographers, with their long history of studying agriculture, food security, vulnerability, and hunger, can and ought to be leading the charge on this scholarly front.

This paper introduces a special issue that critically examines the technocratic and neoliberal agenda for agricultural development and hunger alleviation in Africa. This is an agenda that conceptualises African agriculture that is locally oriented and disarticulated from the global economy as a major problem. Indeed, proponents argue that the existing agricultural model is holding back development, depriving the global economy of needed food stuffs, and exacerbating hunger on the continent (Horta, 2009). The solution to these problems is a 'New Green Revolution for Africa', an approach involving the use of hybrid seeds, fertilisers, and pesticides to boost crop production (Annan, 2007; Toenniessen, Adesina, & DeVries, 2008). It is further argued that this new approach is best adopted within the context of global value chains, that is, African farmers need to be integrated into chains of input suppliers, the processors and transformers of agricultural products, and retailers. This approach involves a new and unprecedented level of public–private partnerships, as donors work to increase the penetration of the private sector and build links between African farmers, input suppliers, agro-dealers, agro-processors, and retailers (Hartmann, 2012). The private sector, in turn, sings the virtues of this new approach and supports it through rhetoric, active involvement in sales, as well as philanthropic activity (e.g. Page, 2012). Within this new regime, hunger is to be addressed in both the rural and urban contexts. Rural hunger is allayed when small farmers produce more crops that they may consume directly or, more importantly, when their higher incomes (achieved through the sale of cash crops) allow them to purchase adequate food. Urban food insecurity is addressed by increased domestic food supply and the penetration of food retailers (mainly of the supermarket format) into poorer neighbourhoods.

The authors in this special issue complicate understandings of this new approach to agricultural development. In The Gambia, Bornstein (2015) adds nuance to the top–down view of seed dissemination by showing that small farmers often demonstrate more agency than previously thought possible when interacting with improved seed varieties such as the much-heralded New Rice for Africa. In contrast, in northern Ghana, Nyantakyi-Frimpong and Bezner-Kerr (2015) find that high external input agriculture is not only ill-suited to the prevailing political economy and ecology of production, but also undermines small farmers' agency in solving day-to-day farming problems. They describe complex and gendered labour dynamics within households that, together with shifting migration patterns, complicate the use of fertiliser and hybrid seed. Peyton, Moseley, and Battersby (2015) take us to Southern Africa where the supermarketization phenomenon is further advanced than in other region of the continent (Reardon & Hopkins, 2007). Here, the Government of South Africa has embraced the penetration of supermarkets into the poorest neighbourhoods as a solution to urban food insecurity. Their analysis shows the limits of this approach as the formalised structure of the supermarket is found to be incompatible with the consumption strategies of the poorest households, revealing the significance of the informal economy in Cape Town and the limitations of the food desert concept[1] for understanding urban food security. Finally, Jones, Schnurr, Carr, and Moseley (2015) reflect on scholarly fieldwork in Africa as it relates to agriculture and food security. They complicate and build upon the assertion that professional development pulls international development practitioners and applied researchers away from the field, offering insightful reflections on the ambiguities associated with maintaining and adapting a commitment to applied fieldwork amid changing professional, personal, and contextual situations.

The twists and turns of African agricultural and food security policy

Today's technocratic and deceptively apolitical agenda for agricultural development and hunger alleviation in Africa is best understood as the latest chapter in a long-standing policy struggle. Agriculture was a priority for many African countries in the years following independence when food self-sufficiency was at the top of the policy agenda (Moseley, Carney, & Becker, 2010). In this era of the 1960s and 1970s, the vast majority of African countries sought to produce as much of their own food as possible in the name of national security. In many cases, African food producers were helped along by heavy investments in agricultural extension, and the introduction of some Green Revolution packages of improved seeds and inputs (especially for maize in Southern Africa and rice in West Africa). The current notion that Africa was 'left out' of the Green Revolution (e.g. Denning et al., 2009) is problematic, given the increased use of external inputs and the turn to Green Revolution approaches in agriculture by many African governments during this period. At this time, food security thinking was similarly premised on the belief that sufficient food supply (a function of home grown production and net imports) was equivalent to food security. As such, the overarching policy of food self-sufficiency neatly tied together agricultural and food security objectives. This is exemplified by the United Nations' approach to measuring food security at the time, known as the food balance approach, which monitored national food supply (domestic production plus net imports) versus the caloric needs of a country (Moseley & Logan, 2005).

The era of food self-sufficiency in African food policy circles came to a crashing halt in the early 1980s with the rise of neoliberalism. The World Bank's (1981) 'Berg Report' stressed rolling back state involvement in development in African countries and giving wider responsibilities to the private sector. Structural adjustment throughout the 1980s and 1990s led to significant retrenchments in agricultural extension and agricultural research (Carney, 2008). As such, there was a dramatic reduction in public development investment in and focus on African agriculture, with a shift away from food crops in this era, and some anaemic interest and investment maintained in cash crops for export such as cotton, cocoa, and flowers (Moseley & Grey, 2008).

While not (necessarily) causally related, conceptions of food security also began to shift dramatically at the same time neoliberal reform was underway in Africa. An important intellectual insight was Amartya Sen's (1981) work on entitlements. Sen argued that hunger was more about inadequate access to food (conceptualised as entitlements) than sufficient quantities of food being available on the market. While fairly straightforward, this conceptual breakthrough was important for understanding a number of examples in African history when the poor were dying of hunger when plenty of food was available. Classic among these cases was the famine which wracked the West African Sahel during the early 1970s (Franke & Chasin, 1980). In this case, people were dying of hunger in Senegal, Mali, and Niger, when peanuts (a key sauce ingredient and source of protein across the region) were being exported to Europe. Further nuance was added to this understanding by scholars like Watts (1983) who, working in northern Nigeria, demonstrated that famine was often a product of social vulnerability created through historical processes rather than drought, overpopulation, or underproduction (see also Watts & Bohle, 1993). In this case, colonialism had fundamentality changed livelihood systems in northern Nigeria, resulting in a shift from subsistence to cash crop production, and a declining practice of storing grain reserves for dry years, and this opened up new spaces of vulnerability.

Sen's conception of food security, with its attention to access, dominated international African food policy circles for about 15 years, from the late 1980s through the early 2000s. While not intended by Sen, his ideas were increasingly enmeshed in a free market approach to food security wherein African countries were encouraged to specialise in a few exports, in which they were deemed to have a comparative advantage, and then trade for food if needed. Free marketers could support the emphasis on 'food access' as long as it meant a focus on poverty alleviation and economic growth. What the neoliberal agenda did not support were investments in social safety nets (as called for by Sen), other than famine early warning systems. This approach seemed to work well throughout the 1980s and 1990s when global food prices were relatively low. Thinking began to shift, however, when global food prices started to slowly rise from 2000 and then spike significantly in 2007–2008 (Moseley et al., 2010).

When the global community's attention shifted back to African agriculture after a 20-year hiatus, the international consensus on 'food security as access' shifted as well. Sen's view of food security succumbed to the rebirth of the technocratic focus on food production as the best way to solve global, and especially African, food insecurity. Beginning in the mid 2000s, momentum started to build among donors for a New Green Revolution in Africa as the best way to address food insecurity. Building on the belief that the first Green Revolution (an international effort to bring hybrid seeds, fertilisers, and pesticides to the Global South) had largely bypassed Africa, the idea was to (re)launch a similar effort more tailored to African crops and conditions (Annan, 2007; Toenniessen et al., 2008). This approach included an emphasis on linking farmers to global value chains (which may terminate in supermarkets for some more developed African urban markets (explored in this issue by Peyton et al., 2015).

Whether or not such an effort will address household food insecurity, as opposed to just increasing food production, depends on the way it is undertaken. McMichael (2013) argued that linking farmers to global value chains will lead to greater financial risks, losses, and debt dependency for smallholder producers (issues alluded to in the northern Ghana study in this issue by Nyantakyi-Frimpong and Bezner-Kerr (2015). Some scholars and advocates argue that a food security-sensitive approach would target the poorest of the poor who are habitually the most food insecure. This means introducing low cost, sustainable enhancements to farming, such as intercropping, agroforestry, composting, and soil conservation measures, in order to help poor, rural households feed themselves yet minimise financial risk. For example, work by Bezner Kerr, Berti, and Shumba, (2010) in northern Malawi has demonstrated that such approaches are possible (see also Patel, Bezner Kerr, Shumba, & Dakishoni, 2014).

Sadly, these types of interventions are only a small portion of the portfolio being pushed by advocates of the New Green Revolution for Africa. Rather, most proponents envision a more capital intensive approach to agriculture involving supply chains of input suppliers, increasingly large producers, agro-processors, and expanding international markets. In some cases, long-term land leases by foreign entities, the so-called land grabs, are also legitimised by this world view as they are seen as a 'show case' for new production strategies. While this approach may improve overall agricultural production in Sub-Saharan Africa, we remain sceptical that it will do much to improve household food security.

How is it that the donor community shifted, in a few short years, away from an entitlements based approach to food security toward one that emphasises food production above all else? The Group of Eight Countries' (G8) New Alliance for Food Security and Nutrition (NAFSN), launched in 2012, is instructive in this regard. This initiative, a

$3 billion commitment by the G8 plus 21 African and 27 multinational companies, aims to lift 50 million people in Africa out of poverty by 2022. This programme is unprecedented in terms of its level of collaboration and coordination with international agribusiness. According to a USAID (n.d.) fact sheet on the initiative: 'The private sector can increase food availability by not only increasing investment in production but also by linking smallholder farmers to broader markets and creating incentives for innovation that improve productivity'. While ostensibly a G8 initiative, the NAFSN is primarily a US led programme. Given the close links between this initiative and the business community, one could argue that those in the American aid circles have learned from the Chinese whose development assistance has very tight links to Chinese business interests. The more likely explanation, however, is the rise of 'philanthrocapitalism' in the US, where former and current business leaders, through the strength of their foundations, have increasingly come to influence the shape and direction of US international development programmes (Thompson, 2014). Central to the philanthrocapitalist world view is a belief that private enterprise is the fundamental agent of progressive change and that business acumen trumps other forms of expertise. It is also very convenient when a company's profit motive lines up nicely with an initiative promoting the end of African hunger (Patel et al., 2014).

Amidst all the fan fare for a New Green Revolution in Africa, the upsurge in donor and corporate interest in African agricultural production, and the rising tide of supermarketization, it is now more important than ever for scholars to take a hard look at the changing African landscape of food production and consumption. The four papers that follow in this special issue open up this research terrain in new and exciting ways, while building on geography's long standing and deep involvement with agriculture and food security issues (National Research Council [NRC], 2010). Geographers' commitment to fieldwork in their study of food and agriculture is also touched on by Jones et al. (2015) in this issue. The reality is that the dominant conception of how best to address food insecurity may have actually regressed under the new constellation of public and private actors working on African agriculture and food security. The problem, of course, is that there is little to no profit to be made in eradicating hunger. The big money, for input providers, agro-processors, and traders, is in building more capital intensive, and market integrated, African farming systems. The renewed technocratic and neoliberal emphasis in agricultural development and poverty reeducation in Africa under the so-called New Green Revolution for Africa, and its supermarket corollary in African cities, may well serve the big money interests, but this collection of initial research on the approach suggests that the extent to which it will enhance household food security for the poorest of the poor remains uncertain.

Note

1. Food deserts are areas where access to nutritious, affordable food is lacking, due to limitations stemming from the geographies of poverty, transportation, and food distribution (Peyton et al., 2015). This idea was first developed in the UK to recognise disparities in people's ability to access healthy food in urban areas (Wrigley, 2002).

Notes on contributors

William Moseley is a professor and chair of Geography, and director of African Studies, at Macalester College. His research interests include political ecology, tropical agriculture, development policy, and food security. Most of his fieldwork has been undertaken in western and southern Africa.

Matthew Schnurr is an associate professor in the Department of International Development Studies at Dalhousie University. He is an environmental geographer with research and teaching interests in environment and development, political ecology, agricultural biotechnology, farmer decision-making, and environmental justice. His work focuses primarily on eastern and southern Africa.

Rachel Bezner Kerr is an associate professor in the Department of Development Sociology at Cornell University. Her research interests focus on farmer-led agroecology, critical examinations of neoliberal approaches to agriculture, and the historical, political, economic, and gender dynamics that shape food security in sub-Saharan Africa.

References

Annan, K. A. (2007, June 14). *Remarks on the launch of the alliance for a green revolution in Africa*. Cape Town: World Economic Forum.

Bezner Kerr, R., Berti, P. R., & Shumba, L. (2010). Effects of a participatory agriculture and nutrition education project on child growth in northern Malawi. *Public Health Nutrition, 14*, 1466–1472.

Bornstein, D. (2015). The social realities of technology transfer: Smallholder farmers' encounter with a new rice variety. *African Geographical Review*. doi:10.1080/19376812.2014.912139

Bourne, J. K. Jr. 2014. The next breadbasket. *National Geographic*. Retrieved from http://www.nationalgeographic.com/foodfeatures/land-grab/

Carney, J. A. (2008). The bitter harvest of Gambian rice policies. *Globalizations, 5*, 129–142.

Denning, G., Kabambe, P., Sanchez, P., Malik, A., Flor, R., Harawa, R., … Sachs, J. (2009). Input subsidies to improve smallholder maize productivity in Malawi: Toward an African Green Revolution. *PLoS Biology, 7*, e1000023.

Franke, R., & Chasin, B. (1980). *Seeds of famine: Ecological destruction and the development dilemma in the West African Sahel*. Montclair: Allanheld, Osmun.

Hartmann, A. (2012). Scaling up agricultural value chains for pro-poor development. In J. F. Linn (Ed.), *Scaling up in agriculture, rural development and nutrition, 2020 vision for food, agriculture, and the environment* (pp. 16–17). Washington, DC: International Food Policy Research Institute.

Horta, L. (2009, May 9). Food security in Africa: China's new rice bowl. *China Brief, 9*, 11.

Jones, J., Schnurr, M. A., Carr, E. R., & Moseley, W. G. (2015). Should I stay or should I go? Incorporating a commitment to fieldwork throughout an academic career. *African Geographical Review*. doi:10.1080/19376812.2014.940986

Mcmichael, P. (2013). Value-chain agriculture and debt relations: Contradictory outcomes. *Third World Quarterly, 34*, 671–690.

Moseley, W. G., Carney, J., & Becker, L. (2010). Neoliberal policy, rural livelihoods and urban food security in West Africa: A comparative study of The Gambia, Côte d'Ivoire and Mali. *Proceedings of the National academy of Sciences of the United States of America, 107*, 5774–5779.

Moseley, W. G., & Gray, L. C. (Eds.). (2008). *Hanging by a thread: Cotton, globalization and poverty in Africa*. Athens, OH: Ohio University Press and Nordic Africa Press.

Moseley, W. G., & Logan, B. I. (2005). Food security. In B. Wisner, C. Toulmin, & R. Chitiga (Eds.), *Toward a new map of Africa* (pp. 133–152). London: Earthscan.

National Research Council. (2010). *Understanding the changing planet: Strategic directions for the geographical sciences*. Washington, DC: National Academy of Sciences of the USA.

Nyantakyi-Frimpong, H., & Bezner Kerr, R. (2015). A political ecology of high-input agriculture in northern Ghana. *African Geographical Review*. doi:10.1080/19376812.2014.929971

Page, G. (2012, August 2). How to ensure the world's food supply. *The Washington Post*. Retrieved from http://www.washingtonpost.com/opinions/how-to-ensure-the-worlds-food-supply/2012/08/02/gJQANPGQSX_story.html

Patel, R. R. Bezner, K., Shumba, L., & Dakishoni, L. (2014). Cook, eat, man, woman: Understanding the new alliance for food security and nutrition, nutritionism and its alternatives from Malawi. *Journal of Peasant Studies*. Retrieved from http://dx.doi.org/10.1080/03066150.2014.971767

Peyton, S., Moseley, W. G., & Battersby, J. (2015). Implications of supermarket expansion on urban food security in Cape Town, South Africa. *African Geographical Review*. doi:10.1080/19376812.2014.1003307

Reardon, T., & Hopkins, R. (2007). The supermarket revolution in developing countries: Policies to address emerging tensions among supermarkets, suppliers and traditional retailers. *European Journal of Development Research, 18*, 522–545.

Sen, A. (1981). *Poverty and famines*. Oxford: Clarendon.

Thompson, C. B. (2014). Philanthrocapitalism: Appropriation of Africa's genetic wealth. *Review of African Political Economy, 41*, 389–405.

Toenniessen, G., Adesina, A., & DeVries, J. (2008). Building an alliance for a green revolution in Africa. *Annals of the New York Academy of Sciences, 1136*, 233–242.

USAID. (n.d.). *More than $3 billion in private sector investment for the new alliance for food security and nutrition*. Washington, DC: US Agency for International Development News & Information. Retrieved from http://www.usaid.gov/news-information/fact-sheets/more-3-billion-private-sector-investment-new-alliance-food-security-and

Watts, M. (1983). *Silent violence: Food, famine and peasantry in Northern Nigeria*. Berkeley, CA: University of California Press.

Watts, M., & Bohle, H. (1993). The space of vulnerability: The causal structure of hunger and famine. *Progress in Human Geography, 17*, 43–67.

World Bank. (1981). *Accelerated development in Sub-Saharan Africa: An Agenda for action*. Washington, DC: Author.

Wrigley, N. (2002). Food deserts in British cities: Policy context and research priorities. *Urban Studies, 39*, 2029–2040.

The social realities of technology transfer: smallholder farmers' encounter with a new rice variety

Daniel Bornstein

Dartmouth College, Hanover, NH, USA

Within the narrative for an African Green Revolution, the continent is often portrayed as the final frontier of global agriculture, the one place where new agricultural technologies have yet to take hold. In West African rice-consuming nations, this technological enthusiasm has revolved around the potential for the New Rice for Africa (NERICA) variety to enable food self-sufficiency. NERICA is an interspecific variety – a cross between the Asian *Oryza sativa* and the African *Oryza glaberrima* varieties – developed by plant breeder Monty Jones (who won the 2004 World Food Prize for this work). What makes the variety unique is that Asian-based traits for high yield were combined with adaptability to African agro-ecological contexts. The Africa Rice Center, part of the Consultative Group on International Agricultural Research, is responsible for its dissemination throughout Africa, a project specifically funded by the African Development Bank and initiated in 2003. In a nod to some of the criticisms directed towards the first Green Revolution of the 1970s, proponents of today's effort tend to explicitly mention the need to target smallholder farmers, whose livelihoods NERICA is aimed at improving.

As NERICA has been rolled out to African smallholders, the response from the more critical perspectives has been to associate NERICA with a broader assault on small-scale farming. Moseley, Carney, and Becker (2010) lump NERICA together with the recent investments in new agricultural technologies, many of which are legitimately a cause for concern in terms of smallholders' autonomy. Yet, this fails to differentiate between those varieties capable of control by farmers and those more likely to erode farmer control. They write, referring to West African governments' increased emphasis on domestic rice production, 'They are assisted in these efforts by a donor community which now seems ready to invest in agriculture again, especially if it is focused on high yielding, high input varieties such as Nerica rice.' In effect, they critique donors' insistence on technological solutions rooted in Malthusian thinking about global food security, which may discredit local knowledge and increase farmers' dependency. The NGO GRAIN (2009) labels NERICA a 'trap for smallholder farmers' and expresses concern about NERICA's association with the 'explosion of private investment in African rice production, which threatens to displace Africa's small-farm rice systems with plantation-style rice production managed by big agribusiness.' GRAIN is particularly alarmed at statements by the high-level officials of the Africa Rice Initiative and Africa Rice Center encouraging private investment in seed production. For example, in Uganda, two seed companies were given exclusive rights to sell the two NERICA varieties most recently introduced into the country (GRAIN, 2009). Furthermore, critics cite the first Green Revolution's tendency to favor large-scale farmers over smallholders, claiming that the new push is failing to learn from this mistake (Holt-Gimenez, Altieri, & Rosset, 2006). The Green Revolution's technological 'package' (ibid.) was most accessible to wealthier large-scale farmers; NERICA's dissemination is embedded within the very institution – the Consultative Group on International Agricultural Research – that is the legacy of the Green Revolution.

This commentary – based on recent fieldwork in The Gambia – draws upon the seed selling and saving practices embodied by smallholder NERICA growers to interrogate portrayals that cast smallholder farmers as passive victims exploited by dissemination of new agricultural technologies. Once these practices – centered on local knowledge, trust, and collaboration – are enacted, they together are a manifestation of exactly the kind of local autonomy that food sovereignty's proponents have been advocating for. The social dimension of NERICA, then, looks awfully similar to instances of seed sovereignty that are altogether an 'offensive, affirmative, positive, proactive undertaking designed to repossess and maintain alternative, (relatively) autonomous spaces' (Kloppenburg, 2010). Seed sovereignty refers to farmers' own control over their access to seeds, independent of external entities; in my study, it specifically encompasses their re-planting of their own seeds and their construction of a system to equitably enable others' access to seed. Farmers' practices around seed saving and selling should be best understood as them asserting their control over the production process, and defying the inevitability of capitalist penetration of agricultural systems.

View from the ground

During a three-month stay in The Gambia in Fall 2012, I witnessed first-hand notions of seed saving and selling. These do not correspond with narratives portraying NERICA as a pernicious threat to smallholder autonomy. Farmers in Jambur – the first village in the country to receive the new variety, in 2006 – have been continuously re-planting their seed ever since, despite National Agricultural Research Institute researchers' recommendation that they plant new seed every 3 years. This guideline, based on agronomic concerns about genetic degradation, constitutes a stopgap measure as the government works toward establishing a formal seed policy. However, farmers in Jambur have continued to employ the same seed selection knowledge that preceded the introduction of NERICA. The president of the local farmers' association argued passionately that the NARI guideline is far detached from an on-the-ground perspective of farmers' experiences, saying it is the farmers themselves who are best positioned to discern which seeds are of high quality. It's unlikely that all farmers were aware of the NARI guideline – after all, the president is the one with the most direct personal contact with NARI and is very well known there – but what's important is that their concrete actions still defy the expected outcome of a new variety's dissemination. The prevalence of farmer-led seed selection over expert knowledge elucidates an important point about the adoption of NERICA: farmers seized on opportunities to continue existing agronomic practices, and in doing so ensured that some localized control over the production process remains intact. These practices were viable because of the inability of NARI to enforce its guideline, and because the seed was not particularly vulnerable to genetic degradation. In this case, then, the skill level of the farmer effectively resisted becoming subordinated to external control over the seed.

Seed selling offers another site to examine this contestation. Some scholarly perspectives, when discussing farmers' autonomy and seed sovereignty, portray farmers as only producers of their own seed, ignoring the reality that farmers also *sell* seed (Stone, 2002). Jambur is widely regarded in Gambia's West Coast region as the village from which to purchase NERICA seeds, in large part because it was the first to receive the new variety, and therefore is well known by people involved in the agriculture sector. Farmers sell their seeds to those from other villages that are either adopting NERICA for the first time, have experienced crop failure, or are looking to expand their NERICA acreage. The executive council of Jambur's village development society sets the prices at which these seeds may be sold. The responsiveness of the council to neighboring farmers' circumstances serves to minimize farmers' dependency on an external seed market. For example, following the 2008 food price spike, the executive council lowered the NERICA seed price from 40 to 30 *dalasies*, helping to facilitate The Gambia's effort to boost domestic production (IRIN, 2008). This move was suggested by the National Agricultural Research Institute, and the council members acknowledged that the price change was the only way to make seed available to other farmers. Also at the prodding of NARI, the price was lowered again, to 25 *dalasies*, in the wake of the 2011 drought that dealt a serious blow to rice harvests.

Though NARI, which is filled with technical experts, may have had an influence in seed pricing, it is worth stressing that fundamentally the seeds themselves still remained under the control of farmers rather than off-farm input dealers, in the sense that the farmers still *produced and sold* the seeds. Interestingly, NARI, by engaging with the farmer-centered seed network, effectively conferred legitimacy on farmer-saved seeds – precisely those seeds harvested in defiance of its own guidelines. In this sense, NARI's involvement should be seen not as constituting outside control over the seed network, but rather as validating the quality of farmer-produced seeds. The existence of this seed market thus affirms that farmer-produced seed can indeed be part of a strategy to boost domestic rice production (Kijima, Otsuka, & Sserunkuuma, 2011), and that access to quality seed does not necessarily require the proliferation of private sector investment. This runs counter to the NARI Director General's view that the NERICA seed sector should be constituted by a value chain that makes seed available from external agro-input dealers, transferring control of the seed away from producers. What might explain NARI's contradiction between its stated objective of formalizing the seed network and its active role in supporting the selling of farmer-saved seeds? While discursively the NARI Director embraced the advent of a commercial seed market based on agro-input dealers, it lacked the capacity to carry out this idea in practice. Such a market would require technical processes for seed certification and an ambitious push for greater presence of private agro-input enterprises. In reality, farmer-to-farmer seed selling appears to be the only mechanism currently capable of delivering seed to farmers who need it. NARI, therefore, offered advice on seed pricing because it had little choice but to engage with the practical realities of the here-and-now, even if it meant supporting a model opposed to its own vision of a formalized seed sector. The farmer-controlled NERICA seed market reflects a hybrid of 'social certification' and 'formal certification' (Sperling & McGuire, 2010): relationships between buyers and sellers – centered on farmer knowledge, fair prices, and trust – effectively combine with a technical entity's influence in the seed sector (Da Via, 2012; Sperling & McGuire, 2010).

Another example of the persistence of farmer-based networks for seeds emerged during the 2011 drought. A farmer from a nearby village, which received no government drought relief in the form of seeds, decided to head to Jambur to replenish his seed stock. This farmer, in fact, was the village's liaison for the government's NERICA dissemination effort – a role that would seem to indicate his close ties to technical experts. Yet it was Jambur, and not the technical experts, that ended up fulfilling his need for new seed in the wake of the drought. This is a reminder of how the farmer-based seed market can become highly necessary in times of crop failure (McGuire, 2007; Sperling & McGuire, 2010).

Taken together, these instances of social certification reveal a very different picture of NERICA than the one painted by critics who claim that the program marginalizes smallholders (GRAIN, 2009; Holt-Gimenez et al., 2006, see above; Moseley et al., 2010). In The Gambia's NERICA seed system, farmers retain some degree of sovereignty over the re-planting and selling their seeds. The dissemination of this new agricultural technology has not resulted in external agents taking over the production process. Recognizing farmer agency is central to grasping the real, lived outcomes of improved varieties. In this case, farmers' deviation from a technocratic model effectively characterizes the social reality of the seed system. Current seed selling and re-planting practices, then, amount to an assertion of an alternative model.

More generally, the case of NERICA reminds us that the rhetoric of agricultural innovation diffusion does not always correspond with the reality on the ground. The triumph of farmers' seed selection knowledge over technical determinations of seed quality testifies to the ingenuity, innovative capacity, and knowledge system embedded in NERICA growers' encounter with this new variety. To conflate agricultural technology transfer with an inexorable commodification of the seed is to overlook the farmer's agency in constructing the social reality of any agricultural system – agency that may enable some degree of farmers' autonomy in the face of the very real structural constraints facing them.

How NERICA can challenge the African Green Revolution model?

What, then, does the case of NERICA teach us about the Green Revolution model more generally? NERICA management practices muddy the waters of the established rhetoric, which critics say is predicated on foreclosing farmer sovereignty over the seed. This case study suggests that – to some extent at least – farmer autonomy over seed access can indeed co-exist with technological change in rice agriculture.

Farmers' quotidian seed selling and saving practices, if allowed to continue, can thus act as a guard against capital's takeover of the seed sector. Their continued use of long-held knowledge about harvesting seeds – dating back to the days when they grew only traditional varieties – has succeeded as a buffer against the imposition of new knowledge forms capable of eroding local control. NERICA's label as an 'improved' variety may, in the eyes of the new GR's critics, portend an inevitable *social* transformation. But farmers' on-the-ground seed management practices in this case look exactly like those that many of the critics are advocating for. Because farmers' control over the seed can be a stepping stone to their sovereignty over the agricultural system more generally (Kloppenburg, 2010), the farmer-led NERICA system has, under certain conditions, the potential to entrench vibrant peasant agriculture.

What are the conditions that enabled some degree of smallholder autonomy to co-exist with technological transfer? Gambian farmers seized on a void left by development actors' incomplete promotion of NERICA: the seed's dissemination occurred without the constructing of an associated enabling environment for reforming the seed sector. In other words, the steps necessary to truly transform the political economy of the seed sector have yet to take hold. There has been little effort to boost NARI's technical capacity to implement the seed certification procedures required for a formal seed market. Private agribusiness investment has not become dominant in the seed sector. No legislation has barred the selling of farmer-saved seeds that lack certification. No law has granted particular (private) entities the exclusive rights to sell NERICA seeds. Many of these features constitute manifestations of the new Green Revolution for Africa. In The Gambia, however, farmer-based seed management strategies have prevailed in the absence of the appropriate enabling environment for private control of rice agriculture.

Acknowledgments

I would like to thank the St. Mary's College of Maryland PEACE Program – particularly Bill Roberts, Yamai Secka-Jack, and Bakary Demba – for all their support during my time in Gambia. Matthew Schnurr and Susannah Chapman provided helpful and constructive feedback on this paper.

Notes on contributor

Daniel Bornstein is a member of the class of 2014 at Dartmouth College, majoring in Geography.

References

Da Via, E. (2012). Seed diversity, farmers' rights, and the politics of re-peasantization. *International Journal of Sociology of Agriculture and Food, 19*, 229–242.

GRAIN. (2009). *NERICA – Another trap for small farmers in Africa*. Retrieved from http://www.grain.org/article/entries/111-nerica-another-trap-for-small-farmers-in-africa

Holt-Gimenez, E., Altieri, M., & Rosset, P. (2006, October). *Ten reasons why the Rockefeller and the Bill and Melinda Gates Foundations' alliance for another Green Revolution will not solve the problems of poverty and hunger in sub-Saharan Africa*. Food First Policy Brief No. 12. Retrieved from www.foodfirst.org

IRIN. (2008, November 6). *Gambia: Can farmers axe rice imports through 'Nerica'?* Retrieved from http://www.irinnews.org/report/81338/gambia-can-farmers-axe-rice-imports-through-nerica

Kijima, Y., Otsuka, K., & Sserunkuuma, D. (2011). An inquiry into constraints on a Green Revolution in sub-Saharan Africa: The case of NERICA rice in Uganda. *World Development, 39*, 77–86.

Kloppenburg, J. (2010). Impeding dispossession, enabling repossession: Biological open source and the recovery of seed sovereignty. *Journal of Agrarian Change, 10*, 367–388.

McGuire, S. (2007). Vulnerability in farmer seed systems: Farmer practices for coping with seed insecurity for *Sorghum* in eastern Ethiopia. *Economic Botany, 61*, 211–222.

Moseley, W. G., Carney, J., & Becker, L. (2010). Neoliberal policy, rural livelihoods, and urban food security in West Africa: A comparative study of The Gambia, Cote d'Ivoire, and Mali. *Proceedings of the National Academy of Sciences, 107*, 5774–5779.

Sperling, L., & McGuire, S. (2010). Understanding and strengthening informal seed markets. *Experimental Agriculture, 46*, 119–136.

Stone, G. (2002). Both sides now: Fallacies in the genetic-modification wars, implications for developing countries, and anthropological perspectives. *Current Anthropology, 43*, 611–630.

A political ecology of high-input agriculture in northern Ghana

Hanson Nyantakyi-Frimpong[a] and Rachel Bezner Kerr[b]

[a]*Department of Geography, The University of Western Ontario, Social Science Centre, London, ON, Canada;* [b]*Department of Development Sociology, Cornell University, Ithaca, NY, USA*

This study traces the trajectory of policy responses to food insecurity in northern Ghana. Historically, the path to agricultural development has been narrowly focused upon deploying technology to increase per capita food production. In the contemporary context, there is a renewed focus on a 'Green Revolution' type of agriculture. Combining village-level fieldwork and geographical perspectives in political ecology, this paper investigates farmer responses to these forms of agricultural intensification. It is argued that input-intensive agriculture is deeply contradictory in the northern Ghanaian context. Agricultural intensification is not only ill-suited to the prevailing political economy and ecology of production, but also undermines small farmers' agency in solving day-to-day farming problems. The findings further reveal how high-input technologies, especially hybrid seeds, are politicized even at the household level of production. From a policy perspective, the findings suggest the strong need to encourage food security initiatives that are sensitive to local context, existing farmer knowledge, and social relations of production. More broadly, the paper contributes to the ongoing debates concerning the form and necessity for a 'new Green Revolution' in Africa.

Introduction

During British colonial rule in the Gold Coast (now Ghana), the colonial administration acknowledged that food insecurity and hunger were severe, persistent and widespread problems in the Northern Territories.[1] Research by scholars and colonial official reports showed severe nutritional deficiencies among residents in the north during the British colonial rule (Cardinall, 1921; Kirk, 1942; Public Records and Archives Administration Department (PRAAD), 1939; Purcell, 1940).[2] For instance, in a treatise by Cardinall (1921) the author wrote that in the Northern Territories, 'harvest is in June and July for early millet [*Eleusine coracana*], and November for guinea-corn and late millet [*Pennisetum glaucum*] ... There is thus a long gap, which is tided over by storing the grain, but is most frequently a period of semi-starvation' (p. 85). In a memoir written in 1942, a British medical officer indicated a telling observation he made when he visited schools in the Lawra area (present day Lawra District). He reported that 'a number of boys showed weaker gain of weight after the holidays, than they did after the term had ended' (Kirk, 1942, p. 42). Apparently, children were in a better position to access nutritious and adequate diets in schools, than they did in their own households.

In another colonial report, an agricultural officer indicated that

> food supplies are often deficient; both in quality and quantity...Because of the distance of these areas from the roads and the relative poverty of the people, the northern savannas do not supplement their nutritional deficiencies by imported food-stuff. Fruits and green vegetables ... are very lacking. (PRAAD, 1939, p. 2)

A dietetic officer made a similar observation when he assessed food and nutrition in the Northern Territories. The officer wrote: 'in the north, nowadays, food is usually available, but very many can ill afford to buy it' (Purcell, 1940, p. 143). The report further showed a marked infant mortality rate of 240 per 1000 births in 1939 (Purcell, 1940, p. 145).[3] For well over half a century after Ghana's political independence in 1957, the grim reality is that hunger has not improved in the northern regions. Arresting evidence of ever-greater food insecurity has recently been inventoried by the United Nations World Food Program (Biederlack & Rivers, 2009; Hjelm & Dasori, 2012).

Against this backdrop, we pose and examine two main questions in this paper. Firstly, what have been governments' policy responses over time to food insecurity in northern Ghana? Secondly, why have these policy initiatives done little to alleviate food insecurity? A critical analysis of these questions is essential in order to examine the sequence in which agri-food policies have evolved, why they did or did not succeed, and what lessons can be drawn for the future. We situate our discussion within the broad analytical lens of political ecology. We take a long-term historical perspective, but also ground our analysis in the contemporary context using evidence from village-level case studies.

Our primary argument is that there is a principal contradiction embodied in government policies aimed at addressing food insecurity in northern Ghana. To date, official responses have largely focused upon agricultural intensification. Such an approach simply recycles the problem as solution. Based upon our village-level case studies, we find evidence suggesting that input intensive technologies are ill-suited to the prevailing political economy and ecology of production. Such farming techniques not only make small farmers vulnerable to market shocks, but also undermine their agency in solving everyday agricultural problems. Furthermore, this approach has adverse implications for intra-household gender relations, engendering tenuous spaces for on-farm labor mobilization. We therefore argue that there is a need for an agricultural approach that draws upon the best of modern technologies and the proven advantages of farmer-led experimentation. Overall, this study contributes to a rich body of scholarship that critically analyzes the role of the colonial and postcolonial state in African communities, the implications of policy for agrarian transformations, and rural communities as repositories of rich knowledge that can be tapped to solve local problems. The paper further contributes to ongoing debates concerning the form and necessity for a 'new Green Revolution' in Africa.

Our discussion proceeds as follows: firstly, we discuss the political ecology framework, then, we trace the historical trajectory of agricultural development policies in Ghana. We then present a case study of maize production technologies, which illustrates farmer perspectives on agricultural technologies, and makes us question some of the assumptions inherent in most Ghanaian agricultural initiatives. We conclude with a commentary on how agricultural policies might be approached differently to improve food security of smallholder households in northern Ghana.

A political ecology analytical approach

We adopt a political ecology approach, which is concerned with environmental, political, and economic processes shaping human–environment relations (Robbins, 2012). Using a historical analysis to understand the current context, we focus on land users and their links to wider environmental and social processes (Zimmerer & Bassett, 2003). We examine social relations of production to illuminate how the microlevel politics of gender and household position influence farmer choices of agricultural technologies. With a focus on the 'land user,' we pay particular attention to how local environmental processes limit or enable different types of farming practices. Our analysis especially bears resonance with, and seeks to advance that of studies that have applied a political ecology framework to analyze input-intensive agriculture (e.g. Carney, 2008; Jarosz, 2012; Weis; 2012; Zimmerer, 2002).

In The Gambia, Carney (2008) has shown how gendered social relations forcefully shape the success or failure of agricultural technologies (see also Carney & Watts, 1991). Zimmerer (2002) has investigated the role of social factors and agro-ecological variability in seed management, revealing how farmer selection practices for traditional and modern seed varieties differed considerably according to socioeconomic status and gender. Watts (2013) used historical and political-economic analysis to explain how food production and famine were shaped by state policies and patterns of surplus extraction in northern Nigeria. This paper builds upon these studies to argue that we must understand the intersection of local environmental practices, power relations and macro-level political economy to see why high levels of food insecurity persist in northern Ghana. In order to advance this argument, we first map out the long-term trajectory of food and agricultural development policies in northern Ghana.

Ghanaian agricultural policies in historical perspective

Ghanaian agricultural development policies have deep colonial roots (Figure 1). Whilst the British established the Gold Coast Colony in 1874, the Northern Territories were not brought under colonial rule until 1902 (Lund, 2003). Initial colonial policies focused upon the production of crops that offered the greatest potential for export to Britain (Seini, 2002). The colonial government paid little attention to the production of non-commercial and staple food crops. Policy emphasis on export crops led to extensive infrastructural development in southern Ghana, where the moist semideciduous vegetation supported the production of cocoa and coffee (Plange, 1979). Very little colonial revenue was spent in the north and the area remained, and still continues to remain, relatively poor and underdeveloped (Al-hassan, 2013; Shepherd, Gyimah-Boadi, Gariba, Plagerson, & Musa, 2004). The only way the north was integrated into the Gold Coast economy was through the provision of labor for southern-based plantation farms (Austin, 2005), which in turn reduced labor availability for food production in the North.

In 1949, the Gold Coast government made the first attempt to improve agriculture in the Northern Territories. As part of a larger and ambitious country-wide program, the government established an Agricultural Development Corporation (ADC). The aim of the ADC was to encourage high-input agriculture in the food crop sector in the Northern Territories, and in the cocoa and rubber subsectors within the southern Gold Coast (Asuming-Brempong, 2003). The ADC was premised on the assumption that indigenous agriculture was inadequate to meet the needs of a growing population and the cash crop

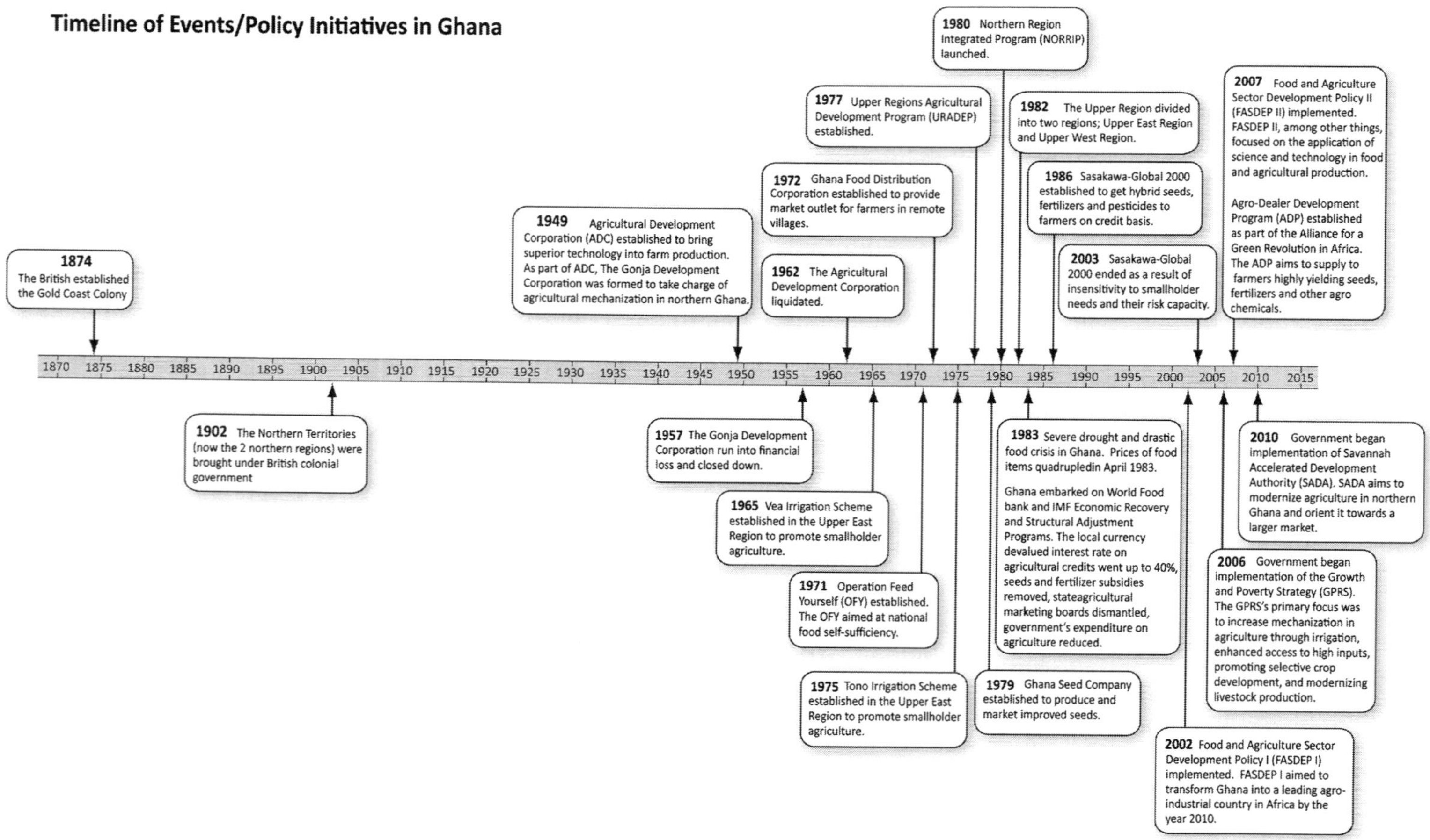

Figure 1. Timeline of major events and agricultural policy initiatives in northern Ghana.
Source: Developed by the authors.

economy in the south. Population pressure was further considered to be a major reason for low-agricultural productivity. The Gonja Development Corporation (GDC) was therefore established as part of the ADC to help ease population pressure and boost food production in the Northern Territories. Initially, the GDC targeted 500 households who were to be resettled from densely populated zones to areas with low-population density (Frimpong-Ansah, 1991). These households were encouraged to go into mechanised cash cropping.

However, the project failed to meet its objectives. Among other reasons, it could not realize the initial target of reaching 500 households, as many farmers refused to leave the presumed overpopulated zones (Hilton, 1960). Further, many farmers refused to adopt the mechanized farming system. One study showed that the 'types of heavy machinery brought in were not suitable for the hard soil conditions; they suffered frequent breakdowns and maintenance was poor. Skilled operators were lacking and the initial land preparation often took away the fertile top soils ... The farmers thought they had better traditional knowledge of farming methods' (Frimpong-Ansah, 1991, p. 83). After independence in 1957, the postcolonial government viewed the ADC as a misguided agricultural initiative. It had accumulated high levels of debt, but made little impact (Hilton, 1960). Both the GDC and the ADC were subsequently closed down in 1957 and 1962, respectively.

The post-independence era saw a number of initiatives to encourage 'modernization' of agriculture in northern Ghana, including encouraging the use of fertilizers, hybrid seeds, and irrigation. In 1965, the government established the Vea Irrigation Scheme, an irrigated area of about 468 ha, to support dry season farming in the Upper East Region. Another large-scale irrigation project, the Tono Irrigation Scheme, was established in 1975 to enhance the diffusion of technology to smallholder farmers in the Upper East Region. In addition, the government established the Ghana Seed Company in 1979 to produce and market improved seeds to farmers. In 1980, the government further initiated the Northern Region Rural Integrated Project (NORRIP) with external funding from the Canadian International Development Agency (CIDA) (Seini & Nyanteng, 2003). The project had several phases including an agricultural component that aimed at diffusing improved technology to small-scale farmers. Although a large number of farmers benefited from the project, it could not be sustained at the end of CIDA's external funding in 1989 (Botchway, 2001).

In the late 1960s and early 1970s, Ghana had a vibrant economy; it was the world's leading producer of cocoa and exported close to 10% of the world's gold (Konadu-Agyemang, 2000). A combination of factors, including government overspending, a fall in cocoa prices, severe droughts, and global oil price increase, led to a complete collapse of the economy in the early 1980s (Pearce, 1992). The government subsequently negotiated for a World Bank and International Monetary Fund Structural Adjustment Program (SAP) in 1983. The SAP came with a loan of US$1.4 billion, with several conditionalities, including deregulation of both input and output markets (Pearce, 1992). The prices of most agricultural chemicals were increased in excess of over 40% per annum, between 1986 and 1992 (Asuming-Brempong, 1994). In 1990, the government eliminated guaranteed minimum prices for food crops such as maize and rice. Fertilizer subsidies were removed in June 1992, while all state marketing boards were also closed down. By August 1992, fertilizer importation and marketing had been privatized (Nyanteng & Seini, 2000).

The effects of structural adjustment were most severe for farmers in northern Ghana where incomes were low (Alderman & Shively, 1996). The removal of tariffs depressed domestic food production by exposing local farmers to a flood of cheap, subsidized rice, and maize from foreign producers. Many smallholders abandoned farming completely, while others switched to export crops that had higher marketing value and received greater credit support (Konadu-Agyemang, 2000). A World Bank report in 1993 confirmed the hardship on northern Ghanaian farmers, stating that 'the deleterious effect [of price liberalization] on incomes (and welfare) may have been more severe among households in northern Ghana where agricultural production is based largely on food crops' (World Bank, 1993, p. 31).

Whilst farmers were still dealing with structural adjustment impacts, an international nongovernmental organization called Sasakawa–Global 2000 (SG-2000) launched its operations in 1986 (Puplampu, 2003). The project aimed 'to bring new agricultural technology to farmers in a rapid and dramatic fashion' (Tripp, 1993,p. 2010). SG-2000 supplied input credit packages to farmers and encouraged them to plant new crop varieties. The project started with 40 plots in 1986; which quickly increased to 1500 in 1987; 15,000 in 1988; and 76,000 in 1989 (Al-Hassan & Poulton, 2009, p. 20). Whilst SG-2000 was not a government initiative per se, it received full governmental support, with government extension agents recruited to assist with the diffusion of technology to farmers (Puplampu, 2003; Tripp, 1993). In northern savannah Ghana, greater emphasis was placed on sorghum and maize.

In the first three years of SG-2000, maize and sorghum yields were remarkable, with farmers recording as high as 40% increase in output (Puplampu, 2003). These results received widespread international press coverage, with the *New York Times* carrying news about 'miraculous seeds' in Ghana (Tripp, 1993). However, SG-2000 proved to be another failure. A key challenge was the issue of credit recovery; according to one report, loan recovery rates fell from 90% in 1987 to 44% in 1989 (Al-Hassan & Poulton, 2009). Moreover, the program was narrowly focused on a few strains of crops (maize and sorghum), and was found to be insensitive to the resource needs and risk capacity of small farmers. When the input credit package ceased as a result of poor loan recovery, all farmers reverted to local seeds and traditional methods of improving soil fertility (Amanor, 2011). The SG-2000 program was therefore closed down in 2003. There is also evidence that SG-2000 produced mixed results in other African countries (Howard, Crawford, Kelly, Demeke, & Jeje, 2003; Keeley & Scoones, 2000).

Since the period after structural adjustment, there has been a renewed focus on high-input agriculture in Ghana. Under different policy initiatives, the government has placed a higher priority on agricultural technologies in order to double yields of farmers, with a focus on northern Ghana (National Development Planning Commission [NDPC], 2005). These initiatives include the Ghana Poverty Reduction Strategy (GPRS), the Food and Agriculture Sector Development Policy (FASDEP), and the Savannah Accelerated Development Authority (SADA). The primary thrust of the GPRS is 'to achieve accelerated growth through modernized agriculture, led by a vibrant and competitive private sector' (NDPC, 2005, p. 29). The SADA program also aims to modernize agriculture in northern Ghana and orient farmers toward a larger market (Al-hassan, 2013). Since 2008, an Agro-Dealer Development Program (ADP) has been launched as part of the Alliance for a Green Revolution in Africa, supported by Rockefeller and the Bill and Melinda Gate Foundations. As part of this initiative, about 2400 agro-dealers have been trained to supply high-yielding seeds and fertilizers to farmers (International Fertilizer Dealer Center, 2011).

After more than half a century of pursuing these agricultural intensification initiatives, there continues to be deepening inequalities between northern and southern Ghana

(Shepherd et al., 2004), while government circles continue to accept high-yielding technologies as the best way forward. In the remainder of this paper, we utilize a case study of maize to investigate whether and how farmers are responding to the mass promotion of input-intensive agriculture. Maize is a useful case study because it is a major staple crop in northern Ghana and the most widely grown cereal in Africa (McCann, 2005). Maize is also tightly linked to the promotion of inputs such as fertilizers, seeds, and pesticides (Gage et al., 2012). In this paper, we use 'landrace' to refer to local seed varieties lacking formal crop improvement. In contrast, we use the term 'hybrids' to refer to cultivars of maize that have been scientifically bred to be uniform and stable, as distinct from landraces (Camacho, Villa Maxted, Scholten, & Ford-Lloyd, 2005).

Research setting

This paper draws upon research in two savannah villages, Jongorro and Hemang,[4] in Ghana's Upper West Region (Figure 2). The Upper West has been one of Ghana's poorest regions for several decades; infrastructure, standard of living, health, and nutritional status are all very low and worse than in any other part of the country (Ghana Statistical Service, 2008). The region falls within the semi-arid, Guinea savannah agro-ecological zone. The vegetation is characterized by a layer of grasses of varying heights, alongside drought-resistant trees. The topography is marked by a relatively flat savannah plains devoted almost exclusively to the cultivation of different kinds of cereals and legumes.

There is a unimodal rainy season lasting for approximately five months, from May to September. The rest of the year is characterized by a pronounced dry season with cold and hazy harmattan weather. The rainfall regime fluctuates considerably between years and within a season. The mean annual rainfall for the period 1953 to 2011 was 1036 mm, with the range varying from a minimum of 523.7 mm in 1986, to a maximum of 1500 mm in 1963 (Figure 3). Dry spells are thus a recurring phenomenon. Rainfall

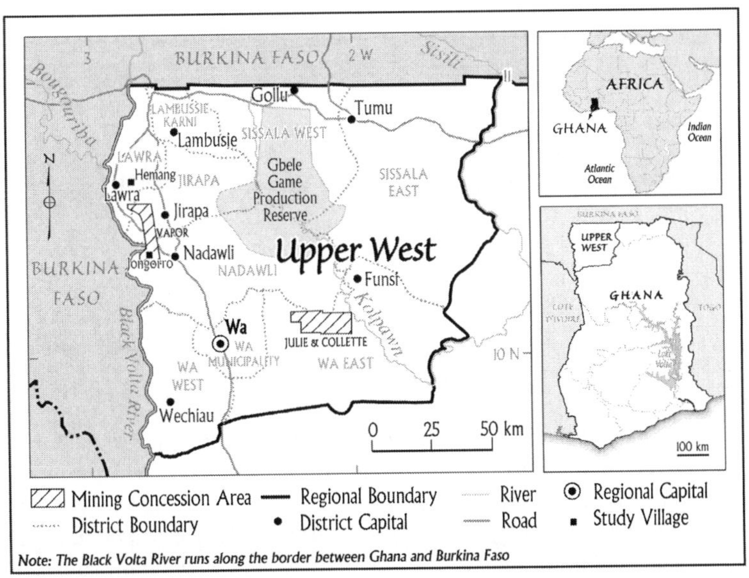

Figure 2. Location of the study villages.

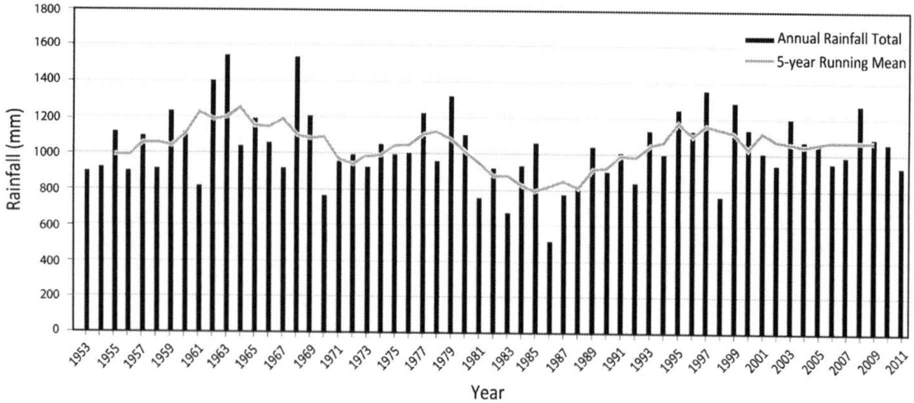

Figure 3. Rainfall variability in Upper-West Region, Ghana.
Data Source: Ghana Methodological Agency, Accra, Ghana.

Table 1. Characteristics of the study villages.

Study Village	Hemang	Jongorro
Geographic location	10°39′43.03″ N, 2°49′21.29″ W	10°21′0″ N, 2°45′0″ W
District	Lawra district	Nadowli district
Population (census 2000)	4041	494
Total households	704	272
Households studied *N* (%)	*N* (35)	*N* (57)
Total houses/farmsteads	305	158
Distance to nearest town	3 km (Lawra)	21 km (Nadawli)
Ethnic composition	Dominated by Dagaabas (95%)	Dominated by Dagaabas (93%)
Electricity	Available	Not available
Market	Not available, 3 km to nearest market	Not available, 12 km to nearest market

Source: Compiled from Ghana Statistical Service (2005); Field notes, January to August, 2012.

tends to occur in heavy torrents and is concentrated in a few days, thereby resulting in heavy erosion, instead of soil moisture recharge (Van der Geest, 2002).

The two research villages are predominantly remote and poor. Table 1 is a summary of site-specific characteristics and differences between the two communities. In both villages, soils are generally poor and thin. Hemang has laterite soils with a moderate acidic content [pH 5.7–6.4] (Adjei-Agyapong & Asiamah, 2002). The same soil type is found in Jongorro, in addition to narrow strips of alluvial soils in the floodplains of the Black Volta River. These alluvial soils are less acidic as compared with the laterites (pH 4.7–5.2), and are extensively used for flood recession farming. Smallholder farming and herding are the principal livelihood activities in the two villages. Production is typically oriented towards household consumption than for market sales. Most households' livestock holdings include cattle, sheep, and goats.

Crops are planted in two types of agricultural fields: the compound field, which often surrounds the homestead, and the bush field, which may be several kilometers away from the village. Fields closest to compounds are intensively cultivated every year. Soil fertility is maintained with manure from livestock kraals and compound sweepings.

Most households cultivate maize, often in addition to pearl millet, guinea corn, groundnuts, and beans. Smallholder farming follows different types of multiple cropping patterns. It may include growing more than one crop on a field during the same farming season (intercropping), growing more than one crop after each other in a sequence (sequential cropping), or growing two or more crops with overlapping cultivation periods (relay cropping). Cereals are intercropped with, or succeeded by legumes, with vegetables cultivated in small patches within the field. This cropping system has a peak planting period from May to June.

Some household members might own individual farms, but a majority of households have a collective field to which junior members (age between 5 and 45 years) and active senior members (age between 46 and 75 years) are expected to contribute labor towards planting, weeding and harvesting. Our survey of time and labor allocation showed that women were more involved in farm operations after initial field preparation. Women's workload usually revolved around weeding, harvesting, and post-harvest work in grain processing. All households had at least one common granary for storing grains harvested from the collective field. At regular periods after harvests, male household heads parcel out grains among the women of the compound. Married women cook for themselves and their children; co-wives alternate the task of cooking for their husbands, although there are some exceptions (in some polygamous households, each co-wife sends a bowl of the evening meal to the husband).

Both villages have significant seasonal migrant populations, most of whom are male. In our surveyed households, about 48% of members in Jongorro and 25% of members in Hemang were absent at the time of our survey. During periods after crop harvest and low-agricultural labor demand, young men and women migrate to work as casual laborers in the forested regions in southern Ghana. These migrant farmers, however, return to their home villages at the beginning of each farming season. The historical root of migration is linked to British colonial policies that largely neglected the north, and post-colonial policies that have further entrenched regional inequality in Ghana (Shepherd et al., 2004). As a result of the large migrant population, food and cash remittances are significant in the local economy (Luginaah et al., 2009).

Methodology

A political ecology approach practically required us to adopt ethnography. As many geographers have emphasized, political ecology's emphasis on 'local knowledge, environmental history, multiscale politics, and socially differentiated resource management practices requires intensive field study and multiple research methods' (Bassett & Zuéli, 2003, p. 117). Our findings are therefore based on ethnographic analysis using a multi-method triangulation technique. From December 2011 to August 2012, the first author collected a series of data (Table 2) and spent a considerable amount of time observing, learning, and interacting with smallholder farmers in order to understand their agricultural practices.

The first author conducted initial oral historical interviews with village elders, before proceeding with an exploratory survey with a random sample of 404 households (Table 2). The survey was subsequently followed by interviews with 18 women and 14 men, and 8 focus group discussions with a total of 75 participants. Following analysis of these data-sets, further interviews were done with 16 women and 12 men, after which theoretical saturation was reached in types of responses (Patton, 2002). Farmers were interviewed inside or near their agricultural fields or homes. Interviews were conducted

Table 2. Methods and data sources.

Methods	Scope and description	Sample size
Questionnaire survey	Closed-ended questions on farm household demographics, food security (using the Household Food Insecurity Access Scale [HFIAS],[c] income, on- and off-farm labor allocation, land holding, and maize variety selection criteria	$N^{\alpha}=249^{a}$, 155^{b}
In-depth interviews	In-depth interviews and observations on household decision-making, division of labor, control over resources, recent crop successes and failures, and livelihood strategies. Analysis of how these characteristics influence the choice of agricultural technologies, especially hybrid maize varieties	$N^{\beta}=28^{a}$, $26^{b,}$ and 6 key informant interviews
Focus group discussions	Semi-structured, gendered, and generational-based focus group discussions on farmers' criteria for maize variety selection. A systematic evaluation of characteristics that farmers considered to be important when selecting seeds	$N^{\beta}=75$ participants in 8 groups
Oral history interviews	In-depth interviews with farmers who were locally known to be skillful cultivators and oral historians. Interviews aimed at identifying local crop taxonomy and landrace diversities present; how landraces were retained in storage from one agricultural cycle to the next; and systems for seed exchange	$N^{\beta} = 5^{a}$, 3^{b}
Farm-based participant observations	Assisted labor in household farms to better understand how field-level agro-ecological conditions (climate, soil properties, weeds, and vegetation) influence the choice of seed varieties and other agricultural technologies	$N^{\beta}=15^{a}$, 15^{b} farms

Note: N^{α} is a random sample; N^{β} is a maximum variation sample.

[a]Total sample in Hemang.

[b]Total sample in Jongorro.

[c]The Household Food Insecurity Access Scale (HFIAS) is a standardized questionnaire and scoring criteria for measuring household food insecurity (Coates et al. (2007)).

in the language preferred by the participant: English, Twi, or Dagaare. The first author conducted all interviews in English and Twi. Research assistants were hired to conduct and interpret interviews in Dagaare. Interviews varied in duration from 3 to 6 h.

We used SPSS to perform statistical analyses. We estimated household food insecurity using a standardized Household Food Security Access Scale (Coates, Swindale, & Blinsky, 2007). Qualitative data from field notes, focus groups, and in-depth interviews were analyzed following the methods outlined by Berg (2004). We used summaries and clustering to reduce the data-sets, and hand-coded interviews and field notes for relevant themes. We recorded all key themes, coding categories, and the number of participants to articulate a particular category. We ensured validity and reliability by using triangulation with multiple methods (Table 2), and presenting preliminary results for verification by research participants (Patton, 2002). Our research methods were approved by the

Non-Medical Research Ethics Board at the University of Western Ontario. In the results section, we have included direct quotations to give voice to participants' own views on agricultural technologies. The quotes were selected based on three criteria: the ability to represent divergent perspectives; typical views expressed by many respondents; and the depth or clarity with which the idea was conveyed.

Research findings and discussion

The surveyed sample had characteristics fairly representative of households in the Upper West Region (Table 3). The mean age of household heads was approximately 56 years and ranged from 21 to 100 years. Household size averaged 7.6 persons in Jongorro and 8 persons in Hemang. However, it was not uncommon to come across larger households composed of a male head (locally referred to as '*landlord*') and his wife or wives; the landlord's unmarried and married sons and their children; the landlord's junior brother (s) – some of whom may be married with one or more wives and several children; and the landlord's unmarried daughters. The average landholding was .6 ha in Jongorro and 2.4 ha in Hemang. The large difference in landholdings is the result of farmland acquisitions for mining in Jongorro.

Whereas hybrid maize has been lauded as high-yielding, pest-resistant, and drought-tolerant, farmers' perceptions were different. The vast majority (76%, $n = 404$) of households planted traditional varieties of maize that farmers' themselves have produced and maintained, often for many generations. Farmers explained that they maintain a large collection of traditional maize seeds, each with unique characteristics to respond to the mix of conditions that unfold in an agricultural season. An older male farmer, with over 45 years of farming experience, stated that 'I plant seeds that I've maintained over the past 30 years. I think the local seeds are more reliable than the store-bought [hybrid] seeds. You can't trust the store-bought seeds. The local maize can produce some grains even in a bad rainfall year.' Similarly, a 65-year-old woman farmer, caring for six

Table 3. Descriptive statistics of surveyed households.

Variable	Jongorro $n = 155$	Hemang $n = 249$	Full Sample $n = 404$	Other studies (Upper West region)[a]
% of household members between 0–14 years	42.8	43.2	43	43.4
% of household members between 15–64 years	50.3	50.6	50.5	50.5
% of household members 65+ years old	6.9	6.2	6.5	6.1
Mean age of household head (years)	58	53.6	55.8	55
Household head never attended school (%)	90.3	80	85.1	69.8
Mean household size	7.6	8	7.8	7.2
Male-headed households (%)	67.7	87.1	77.4	81.7
De facto female-headed households (%)	23	7.2	15.1	18.3
De jure female-headed households (%)	9.3	5.7	7.5	82.1
Households that are Dagaabas (%)	83.6	85.9	84.8	57.5
Mean landholding in hectares	0.6	2.4	1.5	2.7
Households severely food insecure (%)	45	34.5	39.8	34
Households producing maize (%)	100	100	100	100
Households with a migrant in last 2 years (%)	96.8	92.4	94.6	76.3

[a]Biederlack and Rivers (2009, p. 13), Chamberlin (2007, p. 7), Ghana Statistical Service (2005, pp. 88–105), Van der Geest (2002, p. 153).

grandchildren, described her experience this way: 'I've been planting maize nearly all my life, but I've never experimented with the modern varieties on the market. I use seeds that I've kept over the years.'

The large percentage of farmers who exclusively planted maize landrace was in sharp contrast with the percentage of farmers who planted only hybrid maize (8%), or combined both hybrids and landrace (16%) (Table 4). Using the derived scoring system developed by Coates et al. (2007), we identified three categories of food-insecure households in our sample: (mildly food-insecure, 10%; moderately food-insecure, 37%; and severely food-insecure, 39%). Of these three categories, the majority in each group exclusively planted maize landraces. For instance, out of the 156 severely food-insecure households, 117 (75%) exclusively planted maize landrace, while only 14 households (9%) exclusively planted hybrids.

Other studies in Ghana have shown that female-headed households were less likely to adopt improved maize varieties than male-headed households (Doss & Morris, 2001). We investigated whether hybrid maize adoption in our northern Ghanaian sample varied depending on household structure and the gender of household heads. With increasing male out-migration and the growing incidence of female-headed families (Luginaah et al., 2009), simply comparing male- and female-headed households was inadequate. We therefore disaggregated female-headed households into de facto (i.e. the primary household head was a man, but had either migrated, was ill or very old); and *de-jure* (i.e. the primary household head was a single, divorced or widowed woman). This disaggregation was useful in assessing whether women's agricultural decision-making power, especially whilst their husbands are away, influenced the type of technology adopted. In each type of household headship, the majority exclusively planted landrace maize (Figure 4).

During oral historical interviews prior to the household survey (see Table 2), we asked participants how their preferences and circumstances influence their decisions to adopt a particular maize variety. We found that the selection of maize varieties followed a well-defined criterion. The most important considerations included production characteristics such as yield stability, early maturity, drought tolerance, pest resistance, grain weight, cost of seeds, labor as well as fertilizer requirements. A second key consideration was consumption and processing characteristics, including taste, ease of threshing and shelling, storability, and flour-to-grain ratio. These characteristics formed the basis of a set of seed preference survey questions (Table 5).

Table 4. Households' maize production characteristics based on food security status.

Variable	Food secure $n=55$ (14%)	Mildly food insecure $n=42$ (10%)	Moderately food insecure $n=151$ (37%)	Severely food insecure $n=156$ (39%)	All $n=404$ (100%)
Households planting *landrace only*	42 (76%)	36 (86%)	114 (75%)	117 (75%)	309 (76%)
Households planting *hybrid only*	2 (4%)	1 (2%)	14 (9%)	14 (9%)	31 (8%)
Households planting *both landrace and hybrid*	11 (20%)	5 (12%)	23 (15%)	25 (16%)	64 (16%)

Source: Field Survey (2012).

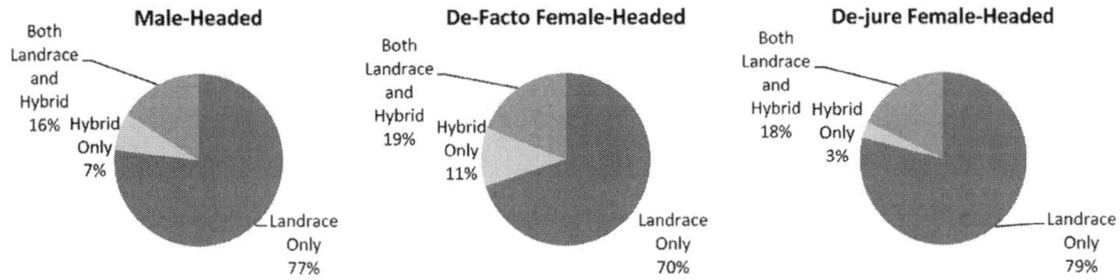

Figure 4. Maize production characteristics based on household headship.
Source: Field Survey, 2012.

Table 5. Maize variety selection characteristics.

	Local varieties			Hybrid varieties		
	All $n=404$	Jongorro $n=155$	Hemang $n=249$	All $n=404$	Jongorro $n=155$	Hemang $n=249$
Production characteristics						
Guaranteed minimum yield	274 (68%)	87 (56%)*	187 (75%)*	130 (32%)	68 (44%)*	62 (25%)*
Early maturity	38 (9%)	12 (8%)	26 (10%)	366 (91%)	143(92%)	223(90%)
Drought resistance	361 (89%)	136 (88%)	225 (90%)	43 (11%)	19 (12%)	24 (10%)
Insect/pest resistance	360 (89%)	141 (91%)	219 (88%)	44 (11%)	14 (9%)	30 (12%)
Lodging resistance	345 (85%)	130 (84%)	215 (86%)	59 (15%)	25 (16%)	34 (14%)
Grain weight	326 (81%)	125 (81%)	201 (81%)	78 (19%)	30 (19%)	48 (19%)
Labor requirement	369 (91%)	141 (91%)	228 (92%)	35 (9%)	14 (9%)	21 (8%)
Cost of seeds	379 (94%)	148 (95%)	231 (93%)	25 (6%)	7 (5%)	18 (7%)
Planting seed availability	382 (95%)	148 (95%)	234 (94%)	22 (5%)	7 (5%)	15 (6%)
Fertilizer requirement	388 (96%)	149 (96%)	239 (96%)	16 (4%)	6 (4%)	10 (4%)
Consumption characteristics						
Taste	386 (96%)	149 (96%)	237 (95%)	18 (4%)	6 (4%)	12 (5%)
Storability	372 (92%)	142 (92%)	230 (92%)	32 (8%)	13 (8%)	19 (8%)
Ease of threshing and shelling	352 (87%)	134 (87%)	218 (88%)	52 (13%)	21 (13%)	31 (12%)
Flour-to-grain ratio	378 (94%)	146 (94%)	232 (93%)	26 (6%)	9 (6%)	17 (7%)

*Chi-square test of homogeneity shows significant differences between Jongorro and Hemang at $p<0.01$.
Source: Field Survey (2012).

For each of the characteristics, we asked farmers if they preferred landrace or hybrid varieties. The majority of households indicated that they plant landrace maize because of characteristics such as guaranteed minimum yield (68%), drought resistance (89%), low-labor requirements (91%), low cost of seeds (94%), higher grain weight (81%), and little or no fertilizer requirement (96%) (Table 5). Furthermore, the vast majority of households indicated that they plant landrace maize because these varieties taste better (96%), are highly resistant to storage pest (92%), are easy to thresh and shell (87%) and have a higher flour-to-grain ratio (94%). Culinary characteristics formed another key factor influencing household decisions to plant landrace, with many farmers remarking that 'local maize dough sticks together better in making Tuozaafi [a thick maize gruel]', whilst others mentioned that 'the dough from local maize does not turn sour as quickly as that from agric [hybrid] maize.' These findings are similar to smallholder farmer concerns in other parts of northern Ghana (Amanor, 2011). The results also support studies indicating that in different parts of sub-Saharan Africa, poorer, food insecure households are more likely to grow landrace maize because of better quality flour, lower cost, yield stability, and higher flour-to-grain ratios (Chirwa, 2005; Lunduka, Fisher, & Snapp, 2012; McCann, 2011).

We performed a comparison test to assess whether the differences between Jongorro and Hemang – for example, soil heterogeneity and distance to input and output markets (Table 1) – will produce variations in farmer responses. Except for one characteristic, that is, yield stability, there were no significant differences in farmer perceptions in both villages (Table 5). Participant observations and in-depth interviews revealed nuanced social dimensions into these survey results. We now turn to a discussion on social relations of production and seed planting ecology to explain *why* farmers select particular maize varieties or allocate them to different growing environments.

Social relations and production politics

> There is an extended work burden associated with hybrid seeds. These seeds make women's work harder! –A Dagaaba woman in interview, July 2012.

> [C]ropping patterns or marketing choices are not the result of a single economic calculus, but are the outcomes of negotiation between husbands and wives, between co-wives and between them and their children. (Scoones, Devereux, & Haddad, 2005, p.3)

During interviews and focus groups, farmers repeatedly stressed that hybrid seed cultivation was a labor-intensive process. Farmers often described hybrid seeds as 'weak,' 'sensitive,' and 'require extra care.' Farmers further emphasized that hybrid maize demand a stricter timing of cultural practices, especially weeding and fertilizer application. For instance, a 69-year-old farmer with three wives and thirteen children, summarized a recurring concern as follows: 'agric seeds [hybrids] should be weeded early in the season, otherwise yield will be poor… This [need for early weeding] increases household labour problems.' Another old farmer (70 years) with two wives and seven children, described his experience by saying:

> agric maize can be good, but not for everyone. It is for the farmer who has something. The farmer who has money. The farmer who has labour. Not every labour, but it means your family members are available and prepared to work harder and carefully. All members should be available to undertake re-planting in case there is a seed failure, or undertake first weeding not later than the first five weeks after planting. The available labour should also be prepared for an additional, second weeding if fertilizer is applied.

Thus, in the annual cropping cycle, the ability of a farmer to mobilize 'a disciplined familial labour force' (Carney & Watts, 1990, p. 231) was a factor that strongly influenced decisions to plant landrace or hybrid maize (see also Moseley, 2000).

Household labor availability was also undermined by the high rates of seasonal outmigration. The complexities and gender dynamics of household migration were such that labor mobilization, coordination, and control could not be feasibly planned. Farmers emphasized that seasonal migrants usually schedule their return around the annual rainfall pattern, an event which in itself was highly unpredictable (Figure 3). Furthermore, men migrants often resisted returning home, enjoying the 'high life' on their own. Women typically tried various means to coax men to return, but were unsuccessful. These negotiations often prevented hybrid seeds from being planted, compelling households to instead plant landrace maize, which as described by one female farmer, 'is sturdy and can withstand delays in weeding without a major lost in yield.' Another interviewed woman who had a migrant husband and children, made a similar observation. She said

> much labour is needed to cultivate hybrid maize as compared to landrace. If I'm sure I can get more labour early in the farming season, I plant hybrids. If not, I plant traditional maize, which can suffer delays in weeding. In the past, we've lost about 70% of our hybrid maize as a result of delays in weeding and fertiliser application.

Another recurrent theme in the interviews was the issue women and junior men's resistance to labor-intensive work routines. Agricultural intensification, using hybrid seeds, was a characteristically coercive process. It often increased the domestic labor demands of certain household members. This configuration is what Carney and Watts (1991, p. 652) have described as 'getting people to work harder, a process that is social and gendered (getting *some* people to work harder than others).' In a culture intimately structured by patriarchal gender relations, we found in particular that hybrids' strict weeding schedules fell upon the shoulders of women, whose domestic work was already overburdened with cleaning, washing, tending the kitchen garden, and fetching firewood and water. 'Our husbands will have nothing to do with weeding the maize plots!' remarked one woman who was very eager to share her story in a focus group meeting. Another woman complained by saying: 'we do most of the weeding while our husbands go drinking. In the evening, the woman has to cook and in the night the woman has to satisfy the man.' Indeed, what was noticeable during field observations was that, in the majority of cases (28 out of 30 households), women and junior men were the locus of crop production after the farm plot had been prepared. Among women, however, juniors, for example, second and third wives, were required to work harder than their senior co-wives.

In some households, these unequal and 'back-breaking work routines', as described by one second wife, have animated struggles over gendered and generational division of labor. As a way of signaling dissatisfaction with strict weeding schedules, some women and junior men engaged in what can be termed 'everyday forms of peasant resistance' (Kerkvliet, 2009). A village headman, for example, lamented about the issue of women and young boys '…not weeding carefully, and deliberately destroying sprouting seedlings.' In another telling account, a 56-year male farmer said: 'the over-reliance on agric [hybrid] maize can cause trouble in the household. Women and other household members can withdraw their labour, leading to insufficient food production.' When asked what he meant by labor withdrawal, he replied by saying 'the women and boys will intentionally delay their return from Techiman [southern Ghana] if they know there will be a lot of hard weeding schedules in the season.' One woman respondent, who was a third wife, explained why women often resist efforts to intensify their labor. She said it 'reduces the hours [they] are able to devote to [their] own vegetable plots, or collect fuel wood for sale.' Eighteen farmers who exclusively planted maize landrace said they do so in order to avoid 'verbal disagreements,' 'hatred,' and 'women quarrels' that are always engendered by the division of farm labor.

The ecology of seed production

Our interview findings showed that compared to hybrids, many farmers preferred landrace maize because of the greater ease to intercrop with groundnuts, cowpea, beans, and bambara nuts. All our interviewees argued passionately that landraces' agronomic practices were fundamentally different from that of hybrids. The traditional maize landrace agronomy involves planting two or three seeds per stand, at regular spacing of about 80 × 40 cm. However, farmers observed that instead of this regular spacing, short-season hybrid maize come with a recommended planting density of 25 × 25 cm row spacing. Interviews and field observations revealed that the higher planting density ensured rapid canopy closure and sunlight interception. Deviations from this recommendation often resulted in considerably lower yields. A majority of farmers however lamented that the higher planting density discouraged intercropping with deep-rooting leguminous crops. Farmers explained that intercropping is only possible when plant population is sparse in order to reduce competition for limited soil moisture at the onset of the rainy season.

One farmer, with over 30 years of farming experience, revealed that 'you have to plant the hybrid maize closer together, which means you can't add beans, groundnut and millet.' Similarly, another farmer stated that 'if you plant agric [hybrid] seed, it is hard to add sorghum, groundnuts and beans.' When asked why he was interested in integrating sorghum, groundnuts and beans into his farming system, the farmer explained that ' …you need to plant those crops if you can't buy fertilizer or get farm labour.' Whilst the farmer did not explicitly employ the language of 'agroecology,' the connection he made reflects leguminous crops' ability to suppress weeds, increase soil porosity, reduce crop pests, and build soil nutrient through nitrogen fixation (Altieri, 2009; Snapp, Blackie, Gilbert, Bezner Kerr, & Kanyama-Phiri, 2010). Indeed, many farmers (33 out of 54) expressed similar reasons why they intercropped maize with legumes.

Twenty-eight farmers said intercropping legumes with maize (either landrace or hybrids) and adding manure provided greater productivity increases than inorganic fertilizers. Studies in other parts of sub-Saharan Africa have found similar impacts from legume intercrops on maize productivity (Snapp et al., 2010). A male farmer, who planted both hybrid and landrace on different plots, was an excellent source of information regarding yield differences between fertilized-maize and maize-legume intercrops. He attempted to compare yields from his two separate maize plots. He explained that he harvested

> five baskets [approximately 150 kg of threshed maize] from the maize-cowpea field, but 3 baskets [approximately 90 kg] from the fertilized-maize field. I plant the local maize here [pointing to a piece of land on his plot] because I don't want to mix it with the hybrid maize, so it keeps pure.

In seven of the eight focus group discussions, participants unanimously agreed that apart from building soil nutrients, intercropping had at least two other benefits. Firstly, farmers said intercropping maize with early millet ensured a quick harvest after the long dry season. These early harvests enable farmers to defray debts and restock depleted granaries. Secondly, farmers argued that intercropping maize with legumes or other cereals insured against production in the face of a variable climate. As one farmer noted 'in case you lose your maize to drought, there is no food to fall back on if you are planting only hybrid maize with no ability to intercrop.'

Field observations and interviews further revealed that landrace varieties were preferred locally because of their yield stability in drought-prone northern Ghana. The most common phrase farmers used to express this view was landrace's 'ability to produce something even in a poor rainfall year.' At the time of this study, hybrid maize seeds were being advertised as short maturing (60–90 days) and drought resistant. However, farmers argued that droughts come at different times in a crop's life cycle, thus, a short-maturing period of 60–90 days did not necessarily imply that hybrids were drought resistant. One male farmer said his long-term farming experience showed that hybrids were unsuited to the agro-climatic conditions in the Upper West Region. He stressed that 'whilst landrace varieties are able to endure droughts and produce cobs under extreme droughts, agric seeds [hybrids] can die or fail to yield if droughts come earlier than anticipated.' Another interviewed middle-age woman expressed similar frustrations with hybrid maize. She said ' ... in 2008, we lost our entire hybrid maize to late season droughts.' An agro-dealer, who was engaged in the sale of hybrid seeds, was asked about his views on some of these farmer concerns. His response was that 'it's a different type of farming and it needs a lot of convincing!'

The cost of seeds and fertilizers

Farmers explained that whereas hybrid seeds must be purchased every season to maintain high yields, local seeds were readily available at no cost. Local farmers relied heavily on informal systems of farmer-to-farmer seed exchanges within and across villages, as well as farmers' own production and careful selection from harvests. In an oral interview with one village elder, who had over 50 years of farming experience, he stressed that informal seed systems have ensured 'access to highly adaptable seed varieties.' During field interactions with farmers, we found women to be at the forefront of seed exchanges. The head of an informal seed exchange group shared this perspective: '... even in exchanging seeds with our friends in other villages, we collect those seeds that will survive in our village environment. We don't exchange any seed. My daughter brought a local maize seed from Wenchi [southern Ghana], but it didn't do well here.'

Furthermore, according to farmers, local maize varieties are less beset by the pest and diseases that affect hybrids. In 14 out of the 30 farms that were visited, farmers noted that although *striga* in landrace can be controlled by using ash and onion residues, the same method could not be used to control *striga* in hybrids. Farmers complained that they needed to purchase agro-chemicals, which were too costly for extensive and timely use. In fact, the cost of seeds was one of the greatest worries facing farmers. Several farmers complained about the ever-increasing seed and fertilizer prices, and the inability to procure credits from financial institutions. In a joint interview with two co-wives, one said 'hybrids produce more yield than landrace'; but the other quickly interjected by saying 'yes, but hybrids also require more fertilizer to produce more cobs.' When asked about the ability to procure loans from financial institutions, farmers often responded by saying 'forget, small farmer, you won't get it.' Nine farmers complained that they had run into debts by borrowing money from friends to purchase seeds, fertilizers and pesticides. A widow with three children narrated the following concern: 'I borrowed money from my neighbours to buy the agric [hybrid] seeds. They all died, so I couldn't pay back.' For a greatest majority of farmers, they planted landrace maize because of the low financial requirements.

Conclusion

In this article, we have examined the historical and current pathways to agricultural development in northern Ghana. Historically, the path to agricultural development has been narrowly focused upon deploying technology to increase food production. Yet, this approach has not produced solutions to benefit smallholder farmers, as food insecurity is still an enduring problem in northern Ghana (Biederlack & Rivers, 2009; Hjelm & Dasori, 2012). Using a political ecology framework, we have examined whether and to what extent farmers are adopting high-input agricultural technologies. In both case study villages, the data suggest that there is low adoption of high-input seed technologies, a finding consistent with studies in different parts of northern Ghana (Amanor, 2011) and elsewhere in sub-Saharan Africa (Lunduka et al., 2012). The analyses further reveal how technology adoption decisions are tightly linked to the coupling of processes operating at a variety of scales. These processes include micro-level gender politics, a semi-arid climate with unreliable rainfall, and large-scale political-economic structures. These political–ecological configurations have produced a context whereby 'technological fixes' entrench structural problems facing smallholders.

The analyses suggest that high-input technologies tend to undermine farmers' agency in solving everyday agricultural problems. An example includes intercropping strategies that improve soil nutrients, reduce crop pest, and ensure crop diversity. Moreover, while hybrid seeds and fertilizer inputs might increase farm productivity, the capital investments are unaffordable in Ghanaian markets already affected by global market restructuring, and domestic market liberalization (Konadu-Agyemang, 2000). At the same time, smallholder access to credit has been firmly curtailed by the lingering impacts of structural adjustment programs. Thus, expensive technologies are being promoted in a system where the local political economy offers little support for the small farmer. The lack of farmer support for the uptake of these technologies raises questions, as Scoones and Thompson (2011) point out, about who benefits, who loses and whose interests are being served with high-input agriculture.

In addition to capital requirements and environmental concerns, our empirical findings point to how gender politics and intra-household labor relations shape maize variety choices. Hybrid seeds impose a radically different labor requirement on households, demanding a firm labor commitment for timely weeding and fertilizer application. Given the customary division of labor by task, age, and gender, female labor power, together with that of junior men, is increasingly used to meet labor requirements. Women and junior men often struggle over labor and resent these intensified work regimes, thereby discouraging increased adoption of hybrid seeds. Farmers contended that the struggles over labor have implications not only for conjugal relations, but create tenuous spaces for labor mobilization in subsequent agricultural seasons. Our case study thus shows how hybrid seeds are politicized even at the household level of production. This finding adds further depth to earlier work in political ecology (e.g. Carney, 2008; Carney & Watts, 1991) suggesting the flawed nature of many agrarian intensification initiatives: that they are labor saving, and that they will particularly benefit women. It is important to emphasize that agrarian intensification should involve not only a simple transfer of technology, but should be sensitive to the local context and social relations of production.

Currently, the socioecological problems facing farmers and their sheer complexities mean that productivity cannot be maximized simply by higher use of technologies. In northern Ghana, many farmers face constraints not directly related to crop yields per se, but severe inequalities in resource access, including land grabbing and unfair trading systems (Laube, Schraven, & Awo 2012; Tsikata & Yaro, 2013). In this sense, a technology-driven approach, and its narrow focus on productivity, may not be a sufficient condition to improve food security. Our argument here is not to undervalue the need for technological investments in agriculture. In the face of severe climatic variability in the region (Laube et al., 2012), it is clear that farmers require additional strategies to build soil health and improve long-term food security. Yet, agricultural intensification on its own will not provide answers to these problems. Rather, there is the need for a multi-functional agricultural approach that values farmer knowledge, considers ecological context, and is sensitive to social inequalities, including class- and gender-based access to resources. Unless these multifaceted issues are carefully considered, promoting input-intensive agriculture will achieve little. A major characteristic of agriculture in northern Ghana is that many soils are poor to start with (Adjei-Agyapong & Asiamah, 2002). Thus, the leaching forces of synthetic fertilizers and other agro-chemicals could have effects on the long-term sustainability of agriculture in the region. Instead of accelerating the use of input-intensive agriculture, policy attention could focus more on supporting diversified farming practices, encouraging on-farm biological diversity, and less use of costly external inputs. Together, such an approach offers greater potential to improve food security and make farming systems more sustainable and resilient (McIntyre, Herren, Wakhungu, & Watson, 2009; Pretty et al., 2006; Snapp et al., 2010).

While our findings are specific to northern Ghana, the broader argument speaks to ongoing debates about the need for an African Green Revolution (Jarosz, 2012; Moseley, 2012). Over the past couple of years, biotechnology, hybrid seeds, and synthetic fertilizers are being promoted as the best and only strategy to address food production deficits in Africa. This exclusive focus on productivity ignores historical experiences that were associated with the Asian Green Revolution, including catastrophic depletion of soils, greater inequalities in incomes, and dramatic decreases in crop diversity (Weis, 2007). This model of agriculture is also insensitive to social differentiation, especially gendered constraints that shape agricultural decisions. As several case studies have shown (e.g. Carney, 2008), gender- and class-based relations are crucial factors influencing who can adopt and benefit from agricultural intensification. In the light of the findings presented in this article, together with that of many others (e.g. Snapp et al., 2010), it is important to rethink the 'New African green Revolution, especially its implications for the long-term sustainability of farming in sub-Saharan Africa.'

Finally, our findings and research approach have relevance for human geographical studies in political ecology. Critical research in this field has explored why neoliberal development interventions often fail to address livelihood needs in specific places, and especially at the local scale (Zimmerer & Bassett, 2003). This article contributes to these studies by demonstrating how micro-politics and social relations contribute to the failures of generic development policies. As the empirical findings illustrate, it is not only agriculture that is being transformed by high-input technologies. So too, are the intra-household social relations. The empirical findings also illustrate the subtle ways in which women and junior men contest and renegotiate the authority of household patriarchs. While these everyday politics and resistance fall short of a broad-based social movement, they can nevertheless be effective in constituting change. Geographers have made numerous calls for political ecology to 'extend the definition of politics from the electoral politics of the state and class to one that includes the political arenas of the household' (Watts & Peet, 2004, p. 33). This paper makes a contribution in this regard. It provides evidence on how state policies conjoin with micro-level gender politics to shape day-to-day agricultural decisions.

Acknowledgements

Karen Van Kerkoerle assisted with the preparation of the map and the timeline graph. We thank Isaac Luginaah, William Moseley and Godfred Boateng for thoughtful comments on earlier drafts of this paper. The paper was originally presented in Los Angeles, California, at an organized session at the 2013 Annual Meeting of the Association of American Geographers. We thank Edward Carr and Matthew Schnurr for their constructive feedback during the session. Finally, we appreciate the useful suggestions from two anonymous reviewers. We are entirely responsible for any remaining shortcomings in interpretation.

Funding

Fieldwork for this research was funded by the International Development Research Centre, Canada; the Centre for International Governance Innovation, Canada; and the University of Western Ontario, Canada.

Notes

1. During the colonial period, the area that was referred to as Northern Territories is now called northern Ghana, and encompasses three administrative regions, namely: Northern Region, Upper East Region and Upper West Region.
2. We do not have archival evidence to consider the pre-colonial conditions of food security, or the role colonial rule played in fostering food insecurity in the Northern Territories.
3. All these narratives are from colonial officers' perspectives. The evidence might therefore be prejudiced, considering that colonial and postcolonial administrators often use exaggerated, misconceived, and crisis narratives to describe African agriculture and landscape change (e.g. see Fairhead & Leach, 1996).
4. Jongorro and Hemang are pseudonyms for the actual villages where fieldwork was conducted.

Notes on contributors

Hanson Nyantakyi-Frimpong is a doctoral candidate in the Department of Geography at the University of Western Ontario. His research interests focus on political ecology, sustainable agriculture, and food security in sub-Saharan Africa.

Rachel Bezner Kerr is an associate professor in the Department of Development Sociology at Cornell University. Her research interests focus on farmer-led agroecology, critical examinations of neoliberal approaches to agriculture, and the historical, political, economic, and gender dynamics that shape food security in sub-Saharan Africa.

References

Adjei-Agyapong, T., & Asiamah, R. D. (2002). *The interim Ghana soil classification system and its relation with the World Reference Base for Soil Resources*. World Soil Resources (Report No. 98). Rome: Food and Agricultural Organization.

Alderman, H., & Shively, G. (1996). Economic reform and food prices: Evidence from markets in Ghana. *World Development, 24*, 521–534.

Al-hassan, S. (2013). Reducing poverty in northern Ghana through the savannah accelerated development authority: What do people expect? In J. A. Yaro (Ed.), *Rural development in northern Ghana* (pp. 249–252). New York, NY: Nova Science.

Al-Hassan, R., & Poulton, C. (2009). *Agriculture and social protection in Ghana*. Future Agriculture Consortium Working Paper SP04.

Altieri, M. A. (2009). Agroecology, small farms, and food sovereignty. *Monthly Review, 61*, 102–113.

Amanor, K. S. (2011). From farmer participation to pro-poor seed markets: The political economy of commercial cereal seed networks in Ghana. *IDS Bulletin, 42*, 48–58.

Asuming-Brempong, S. (1994). Effects of exchange rate liberalization and input-subsidy removal on the competitiveness of cereals in Ghana. In S. A. Barth (Ed.), *Issues in African rural development*, Vol. 2 (pp. 43–59). Arlington, VA: Winrock International.

Asuming-Brempong, S. (2003, October). *Economic and agricultural policy reforms and their effects on the role of agriculture in Ghana*. Paper presented at the Roles of Agriculture International Conference, Rome, Italy.

Austin, G. (2005). *Labour, land, and capital in Ghana: From slavery to free labor in Asante, 1807–1956*. Rochester, NY: University of Rochester Press.

Bassett, T., & Zuéli, K. B. (2003). The Ivorian Savannah: Global narratives and local knowledge of environmental change. In K. S. Zimmerer, & T. J. Bassett (Eds.), *Political ecology: An integrative approach to geography and environment-development studies* (pp. 115–136). New York, NY: Guilford Press.

Berg, B. (2004). *Qualitative research methods for the social sciences* (5th ed.). Boston, MA: Allyn and Bacon.

Biederlack, L., & Rivers, J. (2009). *Comprehensive food security and vulnerability analysis (CFSVA) Ghana*. Rome: United Nations World Food Program.

Botchway, K. (2001). Paradox of empowerment: Reflections on a case study from northern Ghana. *World Development, 29*, 135–153.

Camacho Villa, T. C., Maxted, N., Scholten, M., & Ford-Lloyd, B. (2005). Defining and identifying crop landraces. *Plant Genetic Resources: Characterization and Utilization, 3*, 373–384.

Cardinall, A. W. (1921). *The natives of the Northern Territories of the Gold Coast: Their customs, religion and folklore*. London: Routledge.

Carney, J. (2008). Converting the wetlands, engendering the environment: The intersection of gender with agrarian change in the Gambia. In H. Bauder, & S. Engel-Di Mauro (Eds.), *Critical geographies: A collection of readings* (pp. 402–426). Kelowna, BC: Praxis Press.

Carney, J., & Watts, M. (1990). Manufacturing dissent: Work, gender and the politics of meaning in a peasant society. *Africa: Journal of the International African Institute, 60*, 207–241.

Carney, J., & Watts, M. (1991). Disciplining women? Rice, mechanization, and the evolution of Mandinka gender relations in Senegambia. *Signs: Journal of Women in Culture and Society, 16*, 651–681.

Chamberlin, J. (2007). *Defining smallholder agriculture in Ghana: Who are smallholders, what do they do and how are they linked with markets?* International Food Policy Research Institute (IFPRI) Background Paper (No. GSSP 0006). Washington, DC: IFPRI.

Chirwa, E. W. (2005). Adoption of fertiliser and hybrid seeds by smallholder maize farmers in Southern Malawi. *Development Southern Africa, 22*, 1–12.

Coates, J., Swindale, A., & Blinsky, P. (2007). *Household food insecurity access scale (HFIAS) for measurement of household food access: Indicator guide*, Vol. 3. Washington, DC: USAID.

Doss, C., & Morris, M. L. (2001). How does gender affect the adoption of agricultural technologies? The case of improved maize technology in Ghana. *Agricultural Economics, 25*, 27–39.

Fairhead, J., & Leach, M. (1996). *Misreading the African landscape*. Cambridge: Cambridge University Press.

Frimpong-Ansah, J. H. (1991). *The vampire state in Africa: The political economy of decline in Ghana*. Trenton, NJ: Africa World Press.

Gage, D., Bangnikon, J., Abeka-Afari, H., Hanif, C., Addaquay, J., Antwi, V., & Hale, A. (2012). *The market for maize, rice, soy and warehousing in northern Ghana*. Report on the EAT project, United States Agency for International Development.

Ghana Statistical Service. (2005). *2000 population and housing census: Upper West Region – Analysis of district level data and implications for planning*. Accra: Ghana Statistical Service.

Ghana Statistical Service. (2008). *Ghana living standard survey (GLSS5)*. Accra: Ghana Statistical Service.
Hilton, T. E. (1960). Frafra resettlement and the population problem in Zuarungu. *Bulletin de l'IFAN, XXII*, 426–432.
Hjelm, L., & Dasori, W. (2012). *Comprehensive food security and vulnerability analysis: Focus on northern Ghana*. Accra: World Food Program, Ghana Statistical Service and Ministry of Food and Agriculture.
Howard, J., Crawford, E., Kelly, V., Demeke, M., & Jeje, J. J. (2003). Promoting high-input maize technologies in Africa: The Sasakawa-Global 2000 experience in Ethiopia and Mozambique. *Food Policy, 28*, 335–348.
International Fertilizer Dealer Centre. (2011). *Agro-Dealer development projects increase farmers' access to inputs*. IFDC 2011 Annual Report. Retrieved from http://www.ifdc.org/About/IFDC_Articles/Agro-Dealer_Development_Projects_Increase_Farmers/
Jarosz, L. (2012). Growing inequality: Agricultural revolutions and the political ecology of rural development. *International Journal of Agricultural Sustainability, 10*, 192–199.
Keeley, J., & Scoones, I. (2000). Knowledge, power and politics: The environmental policy-making process in Ethiopia. *The Journal of Modern African Studies, 38*, 89–120.
Kerkvliet, B. J. (2009). Everyday politics in peasant societies (and ours). *The Journal of Peasant Studies, 36*(1), 227–243.
Kirk, J. B. (1942). *Diary of tours in the Gold Coast, as medical officer*. Bodleian Library Archives: University of Oxford.
Konadu-Agyemang, K. (2000). The best of times and the worst of times: Structural adjustment programs and uneven development in Africa: The case of Ghana. *The Professional Geographer, 52*, 469–483.
Laube, W., Schraven, B., & Awo, M. (2012). Smallholder adaptation to climate change: Dynamics and limits in Northern Ghana. *Climatic Change, 111*, 753–774.
Luginaah, I., Weis, T., Galaa, S., Nkrumah, M. K., Benzer-Kerr, R., & Bagah, D. (2009). Environment, migration and food security in the Upper West Region of Ghana. In I. N. Luginaah, & E. K. Yanful (Eds.), *Environment and health in sub-Saharan Africa: Managing an emerging crisis* (pp. 25–38). Dordrecht: Springer.
Lund, C. (2003). Bawku is still volatile: Ethno-political conflict and state recognition in Northern Ghana. *The Journal of Modern African Studies, 41*, 587–610.
Lunduka, R., Fisher, M., & Snapp, S. (2012). Could farmer interest in a diversity of seed attributes explain adoption plateaus for modern maize varieties in Malawi? *Food Policy, 37*, 504–510.
McCann, J. (2005). *Maize and grace: Africa's encounter with a new world crop 1500–2000*. Cambridge, MA: Harvard University Press.
McCann, J. (2011). The political ecology of cereal seed development in Africa: A history of selection. *IDS Bulletin, 42*, 24–35.
McIntyre, B. D., Herren, H. R., Wakhungu, J., & Watson, R. T. (2009). *International assessment of agricultural knowledge, science and technology for development synthesis report: A synthesis of the global and sub-global IAASTD reports*. Washington, DC: Island Press.
Moseley, W. G. (2000). *Paradoxical constraints to agricultural intensification in Malawi: The interplay between labor, land and policy*. Department of Geography Discussion Paper Series (No. 00-1). Athens: University of Georgia.
Moseley, W. G. (2012). Does African agriculture need a Green Revolution? Postscript. In W. G. Moseley (Ed.), *Taking sides: Clashing views on African issues* (pp. 217–228). New York, NY: McGraw-Hill.
National Development Planning Commission (NDPC). (2005). *Growth and poverty reduction strategy (2006–2009)*. Accra: National Development Planning Commission.
Nyanteng, V., & Seini, A. W. (2000). Agricultural policy and the impact on growth and productivity 1970–1995. In E. Aryeetey, J. Harrigan, & M. Nisanke (Eds.), *Economic reforms in Ghana: The miracle and the mirage* (pp. 267–283). Oxford: James Currey.
Patton, M. Q. (2002). *Qualitative research and evaluation methods* (3rd ed.). Thousand Oaks, CA: Sage.
Pearce, R. (1992). Ghana. In A. Duncan, & J. Howell (Eds.), *Structural adjustment and African farmers* (pp. 14–47). London: Overseas Development Institute.

Plange, N.-K. (1979). Underdevelopment in northern Ghana: Natural causes or colonial capitalism? *Review of African Political Economy, 6*, 4–14.
Pretty, J. N., Noble, A. D., Bossio, D., Dixon, J., Hine, R. E., Penning De Vries, F. W. T., & Morison, J. I. L. (2006). Resource-conserving agriculture increases yields in developing countries. *Environmental Science & Technology, 40*, 1114–1119.
Public Records and Archives Administration Department (PRAAD) Ghana. (1939). *Human nutrition report on diet and nutrition in Mamprusi*, December 12, 1939 [ADM 11/1/1294].
Puplampu, K. P. (2003). State-NGO relations and agricultural sector development. In W. Tettey, K. P. Puplampu, & B. Berman (Eds.), *Critical perspectives in politics and socio-economic development in Ghana* (pp. 135–151). Leiden: Brill.
Purcell, F. H. (1940). *Nutrition: Final report on nutrition surveys in the Gold Coast 1940.*
Robbins, P. (2012). *Political ecology: A critical introduction* (2nd ed.). Malden, MA: Wiley-Blackwell.
Scoones, I., Devereux, S., & Haddad, L. (2005). Introduction: New directions for African agriculture. *IDS Bulletin, 36*, 1–12.
Scoones, I., & Thompson, J. (2011). The politics of seed in Africa's Green Revolution: Alternative narratives and competing pathways. *IDS Bulletin, 42*, 1–23.
Seini, A. W. (2002). *Agricultural growth and competitiveness under policy reforms in Ghana.* Institute of Statistical, Social and Economic Research (ISSER) Technical Publication (No. 61). Legon-Accra: ISSER.
Seini, A. W., & Nyanteng, V. K. (2003). *AFRINT macro study: Ghana report (Revised).* Legon-Accra: Institute of Statistical, Social and Economic Research.
Shepherd, A., Gyimah-Boadi, E., Gariba, S., Plagerson, S., & Musa, A. W. (2004). *Bridging the north-south divide in Ghana.* Background Paper for World Development Report 2006.
Snapp, S. S., Blackie, M. J., Gilbert, R. A., Bezner-Kerr, R., & Kanyama-Phiri, G. Y. (2010). Biodiversity can support a greener revolution in Africa. *Proceedings of the National Academy of Sciences, 107*, 20840–20845.
Tripp, R. (1993). Invisible hands, indigenous knowledge and inevitable fads: Challenges to public sector agricultural research in Ghana. *World Development, 21*, 2003–2016.
Tsikata, D., & Yaro, J. A. (2013). When a good business model is not enough: Land transactions and gendered livelihood prospects in rural Ghana. *Feminist Economics, 20*, 202–226.
Van der Geest, K. (2002). *"We're managing!" Vulnerability and response to climate variability and change among rural households in northwest Ghana* (Unpublished M.A. thesis). University of Amsterdam, The Netherlands.
Watts, M. J. (2013). *Silent violence: Food, famine and peasantry in northern Nigeria* (2nd ed. with a New Introduction). Athens, GA: University of Georgia Press.
Watts, M., & Peet, R. (Eds.). (2004). Liberating political ecology. In R. Peet, & M. J. Watts (Eds.), *Liberation ecologies: Environment, development, social movements* (pp. 3–47). London: Routledge.
Weis, T. (2007). *The global food economy: The battle for the future of food.* London: Zed Books.
Weis, T. (2012). A political ecology approach to industrial food production. In M. Koc, J. Sumner, & T. Winson (Eds.), *Critical perspectives in food studies* (pp. 104–121). Toronto: Oxford University Press.
World Bank. (1993). *Ghana 2000 and beyond: Setting the stage for accelerated growth and poverty reduction.* Washington DC: World Bank.
Zimmerer, K. S. (2002). Social and agroecological variability of seed production and the potential collaborative breeding of potatoes in the Andean countries. In D. A. Cleveland, & D. Soleri (Eds.), *Farmers, scientists and plant breeding: Integrating knowledge and practice* (pp. 83–105). Wallingford: CABI Press.
Zimmerer, K. S., & Bassett, T. J. (Eds.). (2003). *Political ecology: An integrative approach to geography and environment-development studies.* New York, NY: Guilford Press.

Implications of supermarket expansion on urban food security in Cape Town, South Africa

Stephen Peyton[a], William G. Moseley[a] and Jane Battersby[b]

[a]Geography Department, Macalester College, MN, USA; [b]Environmental and Geographical Sciences Department, University of Cape Town, African Centre for Cities, Rondebosch, South Africa

The rapid rise in supermarkets in developing countries over the last several decades resulted in radical transformations of food retail systems. In Cape Town, supermarket expansion has coincided with rapid urbanization and food insecurity. In this context, retail modernization has become a powerful market-driven process impacting food access for the poor. The introduction of formal food retail formats is viewed simultaneously as a driver of food accessibility and as a detriment to informal food economies established in lower income neighborhoods. Through a mixed-methods approach, this article assesses the spatial distribution of supermarkets within Cape Town and whether this geography of food retail combats or perpetuates food insecurity, particularly in lower income neighborhoods. Spatial analysis using geographic information systems at a city-wide scale is combined with a qualitative case study utilizing semi-structured interviews and observational analysis in the Philippi township in order to illuminate the limitations of supermarket expansion as a market-oriented alleviation strategy for food insecurity. While supermarkets have been successful in penetrating some low-income communities, they are often incompatible with the consumption strategies of the poorest households, revealing the significance of the informal economy in Cape Town and the limitations of a food desert approach toward understanding urban food security.

1. Introduction

In South Africa, urbanization has been accompanied by a rapid increase in poverty and a rising population of urban poor. Over sixty percent of South Africa's population is now urbanized, and this figure is projected to reach 80% by 2050 (Todes, Kok, Wentzel, Zyl, & Cross, 2010). As a country which still retains elements of its apartheid past in the structural organization of poverty, this trend toward urbanization has exacerbated problems of food insecurity in the poorest neighborhoods, where infrastructure and economic indicators lag behind wealthier regions. In Cape Town, the second largest urban area in the country, with a population approaching four million and a growth rate almost 2% higher than the national average (Battersby, 2011b), urbanization patterns have concentrated the poor in sprawling townships far from the urban core.

South African supermarket expansion has developed in conjunction with neoliberal economic policies promoted by post-apartheid governments. Supermarkets have often been conceptualized as food outlets with high quality, affordable goods, and a wide range of options (e.g. Block & Kouba, 2006; Zenk et al., 2005). In an environment where a structuralized geography of poverty spatially manifests as growing segregated urbanization with limited infrastructural capacity to support it, these supermarkets are conceptualized as market tools to increase food distributional capacity (Reardon & Hopkins, 2006). As such, supermarkets are hypothesized to replace traditional informal markets in lower income areas, leading to an increasingly formalized food system in South Africa (ibid., Reardon, Henson, & Berdegue, 2007). This raises questions concerning the impact of supermarket expansion on people's access to food in urban areas. On the one hand, it appears to be an economically sound market-based strategy backed by the formal sector and government infrastructural development. On the other hand, the informal economy is still a considerable player which many in low-income areas rely on (Battersby, 2011b).

This article utilizes a mixed-methods approach to develop a clearer understanding of the role of supermarkets in either perpetuating or alleviating food insecurity in Cape Town. We argue that although supermarkets have had some success expanding into lower income areas, their capacity to alleviate food insecurity is constrained by their formalized nature, which makes them inaccessible to the lowest income residents. In other words, the supermarket format as a retail outlet, while performing a particular role in low-income neighborhoods, is incapable of meeting all of the food needs of the poorest of the poor. This exposes the limitations of a formal Western market strategy within the hybridized nature of Cape Town's township economies, and the gradual integration between formal and informal economic models. Moreover, it highlights the relevance of a food desert conceptualization of retail access while alternatively emphasizing the need to supplement this with context specific research in food insecure communities.

2. Food deserts as a lens for conceptualizing urban food insecurity

The discourse on food security addresses a diverse set of factors which influence people's access to food. In 1996, the FAO defined food security as 'all people at all times hav[ing] physical and economic access to sufficient, safe, and nutritious food to meet their dietary needs and food preferences for an active and healthy life.' In effect, food security requires more than macro-level food availability, involving an understanding of productive capacity, entitlement failures, structural poverty, variable food access, and cultural constraints.

A prevalent framework for understanding urban food insecurity is the 'food desert' concept, which views food insecurity through the lens of food access. First developed in the United Kingdom to recognize disparities associated with people's ability to access healthy food in urban areas (Wrigley, 2002), this notion is derived from a spatialized epistemology centered around nodes of access in the distribution of food retail. Food deserts are areas where access to nutritious, affordable food is lacking, due to limitations stemming from the geographies of poverty, transportation, and food distribution. There have been many case studies done in both the United Kingdom (i.e. Whelan, Wrigley, Warm, & Cannings, 2002; Wrigley, 2002; Wrigley, Guy, & Lowe, 2002) and the United States (i.e. Block & Kouba, 2006; Larson & Moseley, 2012; Moore & Diez, 2006; Zenk et al., 2005) which have looked at food deserts at the neighborhood scale, with a focus on urban areas.

Studies done on food deserts have consistently emphasized the correlation between poverty, hunger, and food access. Often expressed through the lens of social exclusion, poverty is understood as a 'process of marginalization and deprivation' (Laderchi, Saith, & Stewart, 2003) that excludes people from the socially defined communities in which they live. This means that people who are impoverished not only lack the financial materials necessary for an adequate, comfortable lifestyle, but lack access to the basic social interactions and exchanges which ground them in a community. An approach to understanding poverty should therefore recognize the systemic linkages which reinforce spatially defined patterns of access. The traditional focus on individual households is, while effective for quantifying insecurity, limiting in its reductionism and inflexibility (ibid.).

A large number of definitions for food deserts exist. One definition utilized by the United States Department of Agriculture sees them as 'low-income census tract[s] where either a substantial number or share of residents has low access to a supermarket or large grocery store (ver Ploeg et al., 2009).' Another by Laurence (1997) defines them as 'those areas of inner cities where nutritious food is virtually unobtainable. Car-less residents, unable to reach out-of-town supermarkets, depend on the corner shop where prices are high, products are processed and fresh fruit and vegetables are poor or non-existent.' These definitions indicate the particularly prominent role of supermarkets in this literature as a proxy for access to healthy, affordable food (also see Leete, Bania, & Sparks-Ibanga, 2011). Supermarkets are typically seen as providing a wide range of choices at the most competitive prices, with lower prices than any local stores, fast food restaurants, or gas stations in the area (Block & Kouba, 2006). However, since supermarket distribution is often influenced by market-based factors, they disproportionately locate in higher income neighborhoods to capitalize on a more profitable consumer base. In this way, impoverished communities find supermarkets less accessible, resorting to shopping in stores with fewer healthy choices, lower quality goods, and higher prices, with negative implications for both health and household resiliency (ibid.).

The food desert literature highlights the complex, multifaceted nature of food systems. In urban environments, a multidimensional approach toward understanding food security which combines household scale analysis with a broader understanding of structural determinants must be adopted. In addition to household income and spending capacity, access to retail, and particularly access to supermarkets, is critical for understanding urban food environments.

3. Retail modernization in South Africa

Since the early 2000s, a large body of literature has developed around the recent growth of supermarkets in developing countries (Battersby, 2012; Tschirley, 2007). Much of the early literature on supermarket expansion argued that the introduction of large-scale multinational retail radically transformed local economies in the global South, resulting in food supply chain formalization, conglomeration of small farms, and the exclusion of small local retail through competition and economies of scale (Reardon & Berdegué, 2002; Reardon et al., 2007; Weatherspoon & Reardon, 2003). Small retailers and suppliers in more localized economies were often incapable of matching the low-prices and product diversity of large modern retailers, subsequently going out of business. This in turn resulted in uneven shifts in consumer behavior, benefitting those with the capacity to capitalize on formalization, while negatively impacted the poorest households who had previously relied on these local retailers for their consumption strategies (Reardon et al., 2007).

In South Africa, supermarketization was supported in part by trade liberalization and private development promoted in the African National Congress Growth, Employment, and Redistribution strategy. This strategy is based on a neoliberal premise that development should aspire to 'transformation towards a competitive outward-oriented economy,' (Department of Finance, 1996, p. 1) promotes liberalization through tax-incentives, increased trade and investment flows, and relaxation of foreign exchange controls (ibid., Mubangizi, 2008). This, combined with rising demand from urbanization, led to large-scale expansion of formal food retail in Cape Town, with rapid market saturation in high-income areas (Humphrey, 2007). South Africa-based corporate entities such as Pick n Pay (PnP) and Shoprite were able to rapidly expand, becoming dominant players in the food retail industry. This marked a unique trend in South Africa, where, unlike other African countries which experienced market development through the introduction of foreign supermarkets, the food retail environment was dominated by South African companies. These companies, however, adopted many of the strategies of their Western counterparts, utilizing supply chain formalization and Western-style layouts to establish a ubiquitous supermarket format.

Weatherspoon and Reardon (2003) assess retail modernization and its sweeping influence on food systems in sub-Saharan Africa. The rise of supermarkets in Southern and Eastern Africa marks a shift away from a niche market for the rich toward a factor in lower income household consumption patterns. Some argue that the greater presence of supermarkets in the food economy has negative implications for small farmers and retailers. For example, Weatherspoon and Reardon claim that stores like South Africa's Shoprite and Pick n Pay are requiring farmers to adhere to conditions which support larger scale producers and force small farmers out of competition, resulting in consolidation and industrialization in agricultural production methods. For instance, these supermarkets require larger volume transactions, quality regulations, product standardization, and certifications, all of which require greater costs for the farmer.

Other academics do not see retail modernization as a linear progression from an informal to a formal economy as accompanied by the consolidation and standardization. A more cautious view posits that the historical and cultural significance, as well as the complexity and resiliency of small-scale localized economies, limits this progression. As such, the supermarket revolution while significant is not an all-consuming trend (Abrahams, 2010; Crush & Frayne, 2011; Humphrey, 2007; Strydom, 2011; Tustin & Strydom, 2006). Humphrey (2007) shows that supermarkets have particular difficulty in controlling the fresh food sector. Since controlling decentralized supply systems is much more complicated for fresh food, it becomes difficult for supermarkets to offer lower prices than local stores. Rather than viewing the process as a simple linear progression, Humphrey argues that there are many factors, such as the state of the existing market and trade policies, which influence the format and extent of this transformation.

In South African cities, studies found that rather than a complete transition to buying only from formal retailers, households often utilize a multitude of diverse food and income sources in order to promote resiliency (Crush & Frayne, 2011). Additionally, households reported particular instances when reduced access to food (either from price fluctuations, inconsistent revenue streams, or family illness) would alter consumption patterns, and that they often used informal markets as a coping mechanism (Battersby, 2011b).

The literature on supermarketization and retail modernization provides an alternative framing from that of the food desert literature concerning the role of supermarkets in urban food security. Rather than assuming that supermarket placement corresponds with greater food security, this perspective critiques the influence of supermarkets by taking into account diverse informal economies and the consumption strategies of the poor. Supermarkets are critiqued for destabilizing existing food systems rather than conceptualized as providers of diverse food selection.

4. Methods

This research utilizes a mixed-methods approach to tackle conflicting perspectives in the literature about the role of supermarkets on urban food insecurity in Cape Town. Geographic information systems (GIS) analysis has been utilized to map the spatial distribution of supermarkets and compare it to the geography of poverty and access in the city. This is complemented by a qualitative case study of a neighborhood centered around a supermarket in a township area in Cape Town which was conducted to develop a more in-depth understanding of the role of supermarkets in household food procurement strategies.

Geospatial statistics were utilized to map the distribution of supermarkets and compare this to proxies for food insecurity and accessibility. The study area was determined as the Cape Town municipality, which consists of 24 sub-councils comprising 111 wards (City of Cape Town, 2012). Since census data are not available at the ward scale, sub-place data, which provides a smaller unit of analysis (with a total of 683 sub-places in the Cape Town municipality) was used. GIS was used to map and analyze supermarket location against sub-place scale household-level data.

Considering the market dominance by Shoprite, PnP, Woolworths, and Spar (Weatherspoon & Reardon, 2003) this research looks specifically at the spatial distribution of these four chains throughout the greater Cape Town metropolitan area. Supermarket locations were gathered from online store locators on the websites of Shoprite, PnP, Woolworths, and Spar, respectively.

Food insecurity is a complex, multifaceted problem with a wide array of determinants ranging from price fluctuations in global markets to vulnerability to disease. Unfortunately, an assessment of all variables which influence food insecurity is beyond the scope of this study. Recognizing this limitation, average annual household income was utilized as a proxy for food insecurity because there is a well-documented, direct relationship between the two variables (Battersby, 2012; Sen, 1982). This variable was assessed against supermarket distribution. In Cape Town's urban environments, food production is not necessarily a viable option and is rarely practiced to promote resiliency in food security (Battersby, 2011a; Crush & Frayne, 2011; Maxwell, 1999). Where the majority of foodstuffs are obtained through cash-based transactions, household income is a crucial component in food security conversations. Although not fully representative of food insecurity, household income is utilized for its ability to illuminate variation among sub-places.

Data on average annual household income was taken from 2001 city census data.[1] A quintile breaks method was utilized to divide up the data into five categories with an equal number of values, as the high positive skew in the data's distribution would cause both equal interval and optimal breaks classification methods to over-generalize the variations among low-income sub-places. Outliers found more than two standard deviations above the mean were removed.

In addition to mapping supermarket distribution, qualitative field research was performed around the shopping center located at the corner of New Eisleben and Landsdowne Roads (see Figure 3) in order to understand the relationship between formal and informal food service formats. This location was chosen for two reasons. First, it is one of the few sites in Cape Town in which case study data on the extent of food security is present. A survey conducted by the African Food Security Urban Network found 17 percent of residents in Ward 34 to be moderately food insecure, and 71 percent to be severely food insecure (Battersby, 2011b). Second, it has a prominent presence of an informal economy. This is relevant because literature on supermarket diffusion has debated over the role of the informal economy in relation to the expansion of supermarkets (i.e. Abrahams, 2010; D'Haese & Van Huylenbroeck, 2005; Humphrey, 2007).

Semi-structured interviews were conducted with owners and employers of three different informal retail formats. Judgment (or purposeful) sampling was utilized, in which sample participants were chosen through available knowledge of variability and participant significance as derived from discussions with a translator prior to field work (Marshall, 1996). Maximum variation and snowball sampling shaped the total sample size, with data saturation determining the sample size at 20 (Hay, 2010).[2]

For the purposes of this research, the informal retail formats that were analyzed will be defined here. Fruit and vegetable (F&V) stands are retailers that present their produce on land which they lack ownership to, in an open-air format with limited infrastructure. They sell predominantly fruits and vegetables, and are typically operated by one to three workers. Meat stands also operate on public land, but with higher levels of infrastructure. They typically have the capacity to cook their meat and house a limited number of customers at their stands. The size and scale of distribution among meat stands varied considerably. Finally, Spaza shops are stores which sell commonly purchased products, including packaged staple foods, cigarettes, and 'air-time' (phone credits), which are run out of informal dwellings in an area regulated as residential. Spaza shops also vary widely in their size and business capacity.

5. Supermarket distribution in Cape Town

The distribution of supermarkets was compared with average annual household income to determine whether a relationship exists between the two variables. The number of supermarkets per square kilometer and also per 1000 households was calculated per income category. The income categories were: R1,454.62 to R43,197.89, R43,261.85 to R78,504.71, R78,521.88 to R122,188.25, R122,418.13 to R188,895.03, and R188,914.92 to R350,711.77, listed from 1 to 5 ascending in value.[3]

This study found that the distribution of supermarkets follows spatial patterns of uneven dispersion throughout the city. As indicated in Figure 1, the majority of supermarkets are located in the southern suburbs, northern suburbs, City Bowl, and Atlantic Seaboard. Those areas with fewer supermarkets are in the Cape Flats townships as well as the spectrum of low-density rural to peri-urban landscapes found north of the Northern Suburbs.

Findings from the analysis of supermarkets per square kilometer found that supermarkets were most common in middle-class neighborhoods, with income group 4 having .338 supermarkets per square kilometer (see Figure 1). This value declines in income group 5 to a rate of .164 supermarkets per kilometer. However, it is the lower three income groups, and particularly group 1, encompassing the lowest income range that faces the lowest ratios of supermarkets per square kilometer. Income group 1 contained .021 supermarkets per kilometer, followed by a rate of .096 supermarkets per kilometer

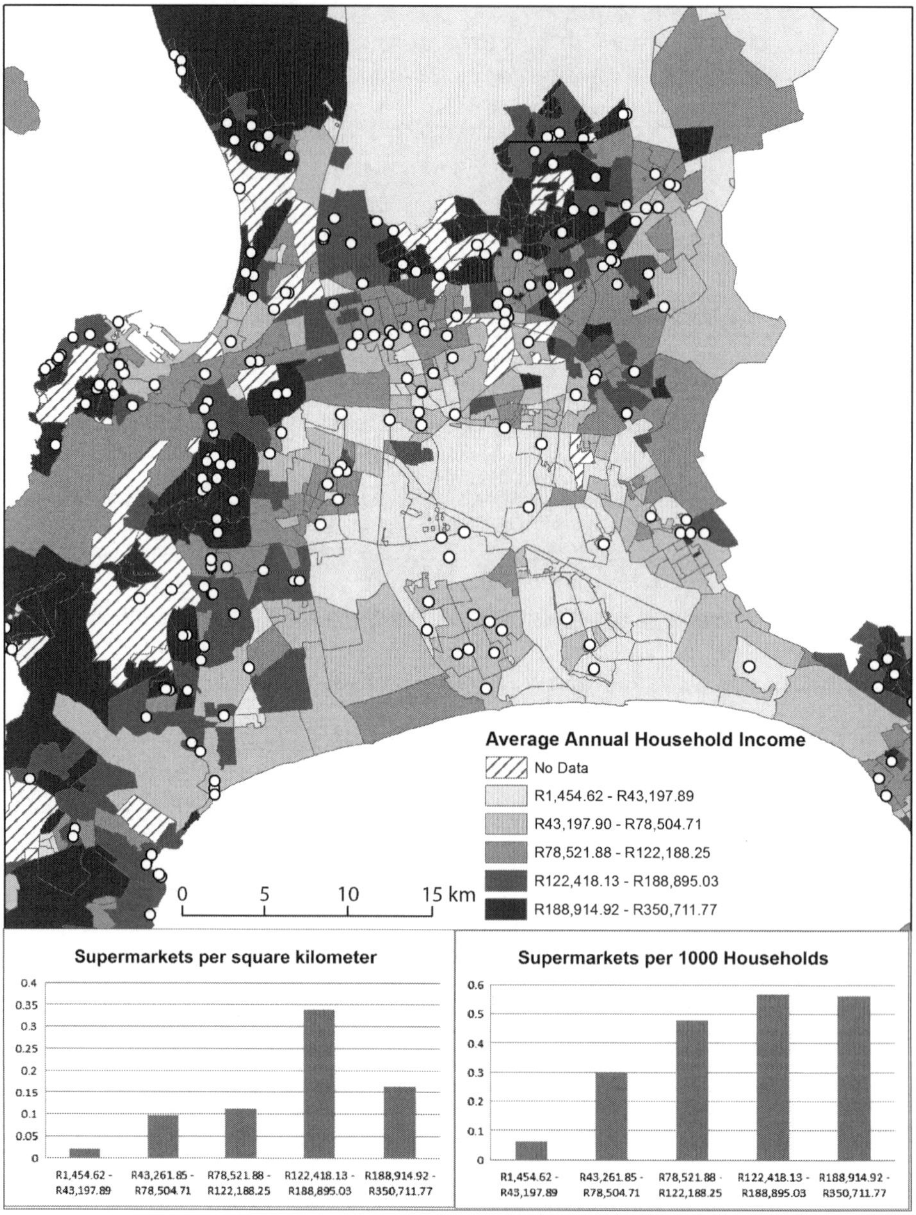

Figure 1. Supermarket distribution in Cape Town.
Notes: Map of supermarket locations (points) in relation to average annual household income (choropleth map), with corresponding graphs representing density measurements of supermarkets by area and number of households.
Cartography by first author (Stephen Peyton), data from 2001 census.

in category 2 and .112 supermarkets per square kilometer in group 3. When comparing income groups 4 and 1, you find that the supermarket density of group 4 is more than sixteen times the supermarket density of category 1.

Assessment of supermarkets per 1000 households also found discrepancies between income groups, but portrayed a different relationship between groups (see Figure 1). Group 4 was again found to contain the largest ratio of supermarkets to households, with a value of .567 supermarkets to every 1000 households. However, unlike the previous density measurement, group 5 was relatively similar, with .561 supermarkets per 1000 households. Additionally, group 3 appeared to have a relatively high ratio, with .477 supermarkets per 1000 households. On the other hand, the lowest income groups displayed a dearth of supermarkets, with .063 and .297 supermarkets to every 1000 households, respectively. The extremely low value of group 1 in particular should be noted, as group 4 has more than nine times the number of supermarkets per 1000 households.

The discrepancies between these two sets of data lie in the variability of household density[4] within sub-places and the market-based factors which influence supermarket placement. Differences in the relationship between groups 4 and 5 in the two data-sets indicate that after a certain threshold, the density of supermarkets begins to decline, likely due to the reduced profitability of supermarkets in high property value neighborhoods. However, since the density of residents in these neighborhoods also declines, the ratio of supermarkets to households stays relatively constant. The differences in income between groups 2 and 3 likely correspond to a decrease in the density of household settlement between group 2 and 3. Although the density of supermarkets themselves is not changing by much between income groups 2 and 3, the density of households is, resulting in higher values for the latter analysis. However, a more constructive conceptualization of this comparison involves recognition of the high-density settlement in the townships corresponding to the first two categories.

These relationships highlight the unequal distribution of supermarkets in Cape Town by average annual household income. In line with the food desert literature above, the spatial organization of supermarkets indicates access limitations for the poorest households, with limited success in expansion to the poorest townships. However, attempts have been made by large-scale food retailers to capture lower income consumers. For example USave, a Shoprite-owned retail format with low-price bulk goods, was developed specifically to target the poor. On their website, Shoprite claims that 'The USave chain's focus is the lower income groups. Customers who are serious about saving and do not need the expensive frills and spills of regular shopping centers, are invited to put USave to the test (see website, www.shopriteholdings.co.za).'

The distribution of USaves differed drastically from the distribution of supermarkets as a whole. USaves tended to locate more frequently in the lower income Cape Flats area, rather than the higher income (predominantly white) suburbs and CBD. However, their distribution in low-income townships has been limited mainly to the edges of the Cape Flats region (see Figure 2). This has provided many in lower income neighborhoods with a cheaper alternative food source, but it has neglected those most in need; those in the central Cape Flats region, where poverty is most heavily concentrated.

Income group 2 held the highest values for both USaves per square kilometer and USaves per 1000 households, at .0181 and .0558, respectively (see Figure 2). Conversely, groups 3, 4, and 5 had relatively low values in both equations, with .0046, .0057, and .0036 USaves per square kilometer and .0195, .0096, and .0122 USaves per 1000 households, respectively. This indicates a particular targeting strategy in which those in the lower income range of income group 2 appeared to have benefitted greatly. Furthermore, the severe drop in supermarkets in the higher income groups indicates that

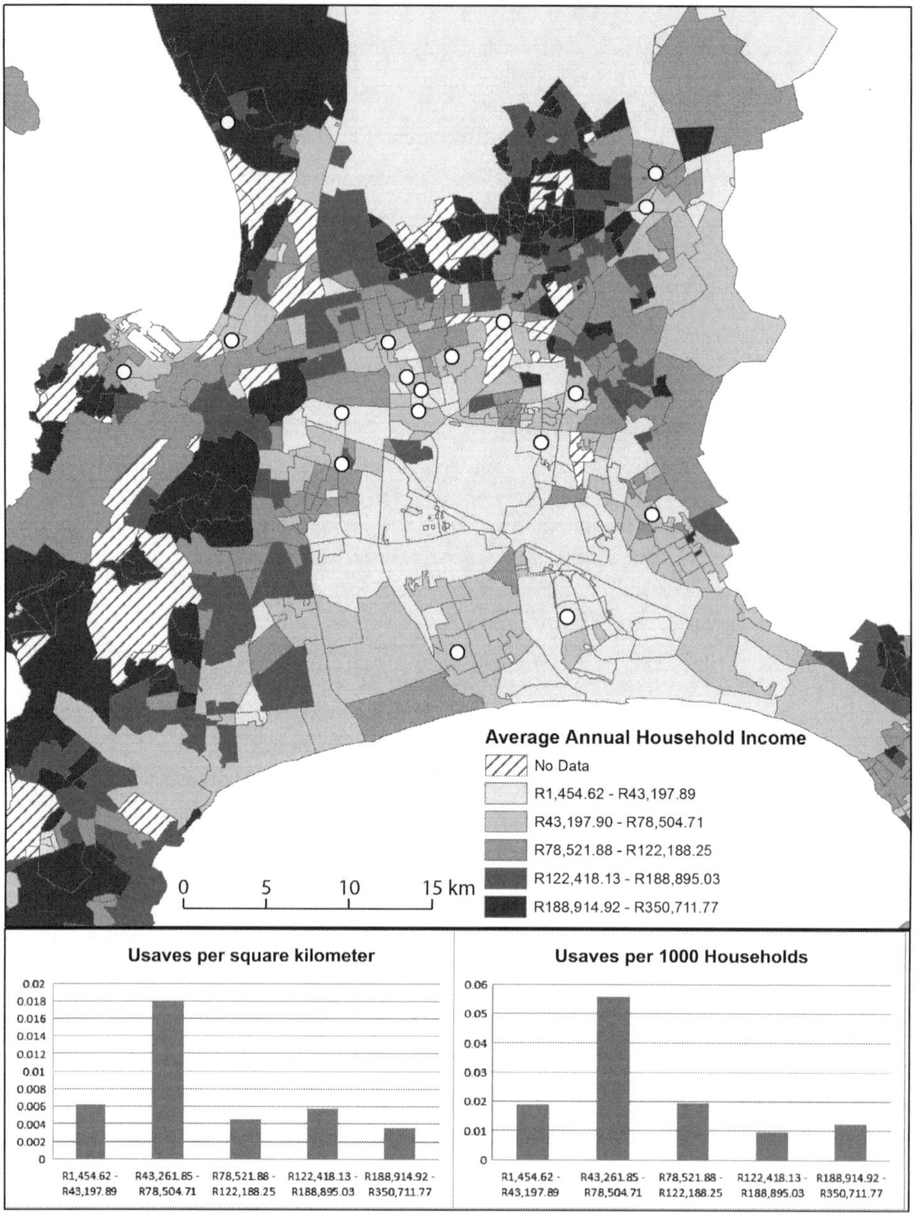

Figure 2. Usave distribution in Cape Town.
Notes: Map of Usave locations (points) in relation to average annual household income (choropleth map), with corresponding graphs representing density measurements of supermarkets by area and number of households.
Cartography by first author (Stephen Peyton), data from 2001 census.

this retail format is less compatible with the shopping preferences of higher income households and therefore only minimally profitable in higher income sub-places.

This targeted market strategy has failed to fully diffuse into sub-places facing the greatest impacts of structural poverty. Income group 1 only hosted .0062 USaves per square kilometer and .0189 USaves per 1000 households. Although these values are higher than those recorded for groups 3 to 5, they still represent values considerably lower than those found in group 2, with its values being almost three times as high in both cases. This trend, mirroring the evidence seen above, reveals the limitations of supermarkets in penetrating low-income neighborhoods, implying a structural disconnect between the needs of the poor and the strategies of the retailer. Even when private investment is directed toward lower income communities (such as with the USave example above), it appears the poorest of the poor fails to be recognized as an adequate consumer base.

These findings highlight the relationship between average annual household income and supermarket prevalence, revealing the constraints facing supermarket diffusion into low-income environments in Cape Town today. Even in the case of ventures specifically targeting lower income communities, supermarketization comes up short. Due to structural impediments to social inclusion for communities facing poverty, the formal food retail system appears to have been inadequate in addressing the food needs of the poor.

These findings build off of growing concerns voiced by AFSUN in their State of Food Insecurity report for Cape Town (Battersby, 2011a). Poverty has been recognized as an increasing problem in the quickly urbanizing metropolitan region and food insecurity is prevalent in many areas of the city (ibid.). These findings indicate that in this environment of structural inequality and an associated geography of poverty, the market-based expansion of supermarkets is limited in its capacity to tackle food insecurity. The Cape Flats region, recognized as highly food insecure in the 2008 AFSUN survey, has attracted minimal supermarket development, which tends to spatially organize around the periphery of the township conglomeration. Conversely, high income areas in the Northern and Southern Suburbs and the CBD have enjoyed drastically higher supermarket prevalence, with 10 to 40 times the frequency of supermarkets either per unit area or per household.

In the context of traditional food desert literature, the inequalities in supermarket access reveal the prevalence of food deserts in the geography of food retail in Cape Town. The lack of supermarkets indicates a reduced capacity to access healthy, diverse food choices in low-income townships. However, while supermarkets aren't as prevalent in lower income areas of Cape Town, in order to truly understand their impact on food insecurity, it is important to examine their current role in relation to alternative food sources in low-income areas where they do exist, and how these relationships illuminate both the benefits and limitations of supermarket expansion for food insecure households.

6. Interactions between formal and informal food markets

A case study was conducted looking at informal business strategies and household consumption patterns in a neighborhood situated near a supermarket in the Philippi area of the Cape Flats. Philippi is bordered by Nyanga and Gugulethu to the north and Lentegeur and Weltevreden Valley to the South. Found within the Mitchell's Plain main-place, Philippi has a relatively low average household income, within both the Cape Flats specifically and Cape Town as a whole (see Figure 3). This is reflected in the fact that while Mitchell's Plain residents had an average annual household income of approximately R41,560 at the time of the 2001 census, Philippi residents only held an average annual household income of roughly R14,774. This contrast is even greater when compared to the average annual household income of the Cape Town metropolitan area as a whole, which was approximately R80,957. Philippi also had a relatively high ratio of informal to formal dwellings, with 55.13% of dwellings being labeled as informal,[5] and

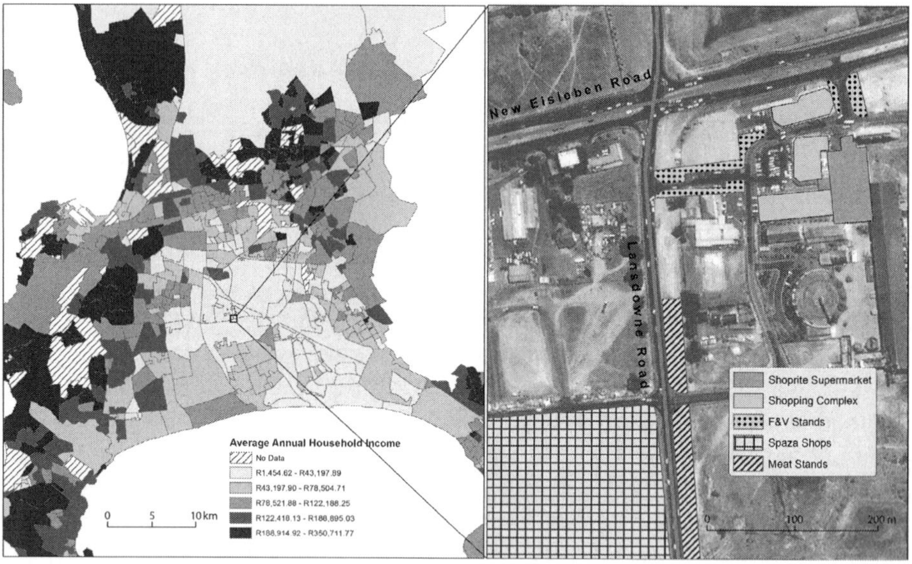

Figure 3. Reference map of case study research area.
Notes: Map of location where qualitative research took place. Case study detailing location of different store types, referenced as an inset map in relation to the larger Cape Town metropolitan area as denoted by a choropleth map of average annual household income.
Cartography by first author (Stephen Peyton), data from 2001 census.

39.29% being labeled as formal (other houses were labeled under different formats). On the other hand, Mitchell's Plain as a whole had a ratio of 21.64% informal dwellings to 72.87% formal dwellings. Dwelling composition as seen in the entire Cape Town municipality is composed of 73.86% formal and 18.4% informal dwellings, also indicating the massive discrepancies between sub-places and the particularly informal nature of the Philippi area.

As noted previously, this research was conducted around a Shoprite supermarket located at the intersection of New Eisleben and Lansdowne Roads. Housed within a shopping center, this supermarket performs the role of an anchor store, supported by several small clothing and fast food retailers housed within the complex. Surrounding the shopping center are several forms of informal food retail (see Figure 3). Immediately surrounding the shopping center, within the parking lot of the development are fruit and vegetable stands (F&V stands). Further out, along Lansdowne Road, a prominent transportation artery, are a row of meat stands. Further away still, Spaza shops are found interspersed among informal residential units. This research location was chosen for its relevance to food security data collected during the 2008 AFSUN survey and the interesting interplay between formal and informal food retail formats.

The informal economy operates where the market-based strategies of formal institutions and organizations are incapable of meeting consumer demand due to incompatibilities between formalized systems and households facing poverty. In Philippi, this research found that due to their informal nature, retailers were able to capitalize on consumers with inconsistent income streams and low disposable income. Informal retail business was found to be a coping mechanism to make up for the lack of formal employment opportunities.

In order to develop a strong consumer base, many informal retailers worked on a credit system. The majority of Spaza shops interviewed claimed to provide goods on credit to regular customers, with the assumption that they would be paid for at a later date. Concerning informal credit-based systems, one Spaza shop owner stated:

> You see, sometimes I do [give credit], if somebody come here and tell me, 'I have this problem, help me with this thing,' and the thing, I have to help him with because as long as I know him, I help. At the end of the day, he'll help me back. (Interview with respondent #9, 6/29/2012)

As seen in the quote above, this informal safety net system, which expands beyond monetary credit to include favors, is predicated on the existence of strong social ties. These exchanges require loyalty from the consumer as an informal socially binding contract which develops a mutually beneficial relationship. These informal relationships are vital, since competition solely through the pricing of products is often ineffective, considering the limitations of consumer spendable income. As such, factors such as product price, product quality, customer interactions, and relationships are weighted differently in informal business practices. Consumers would often decide to shop consistently at one particular store, and shopping at an alternative would be heavily frowned upon. One meat stand owner said this about his relationship to his customers:

> We [should be] clocking out at 7 but we are clocking out at 9, because there are a lot of customers, and they are complaining 'no, why you didn't make my meat, you can't close without me,' and you, you can't close, you know. (Interview with respondent #20, 6/29/2012)

Although each retail type had different spatial and economic organization, they all shared a common trait in their relationship to the supermarket. None of those interviewed expressed any concern of competition from the supermarket; competition was primarily between retailers of the same informal format. In fact, both F&V and meat stands actually capitalized on the existence of the shopping center through their spatial organization.

F&V stands were found most proximate to the shopping center, deliberately positioning themselves to capture the consumer base attracted by the supermarket. The supermarket is seen as a central location, with a high gravitational potential, bringing in customers who may support alternative businesses located close by. Most participants responded that their main consumer base was those who had bought from Shoprite but had forgotten to purchase a particular product. As one F&V stand worker stated,

> [Shoprite] customers, sometimes they forget to buy something from Shoprite, and then when they pass through the road that person's mind was to buy [fruit and veg] from Shoprite not from here but because he's far from Shoprite now … this stand is here because this is where everybody is going through and you can't put a stand where nobody is going. (Interview with respondent #3, 6/20/2012)

Meat stands were also found to capitalize on the centrality of the supermarket, utilizing the high level of traffic along Lansdowne Road, an artery which runs directly through the Cape Flats into Khayelitsha and connects the townships to the Southern suburbs. This was a deliberate decision in order to generate a larger potential consumer base. One participant stated that the shop was opened in that particular location because

> it is near to the main road. This is a main road here. So it was all the people across Philippi can see the meat place. (Interview with respondent #18, 6/29/2012)

These retailers were found to have developed what they perceived to be a mutualistic relationship with the Shoprite. One worker of one of the two largest stands said in regards to competition with Shoprite:

> No, it can't hurt our business, because we are not selling the same thing. We are selling a different thing. So maybe a person can come here to buy meat but he thinks also, I must buy something at Shoprite. So a person can come to Shoprite and think oh I can go buy the meat as well. (Interview with respondent #20, 6/29/2012)

Rather than a purely parasitic relationship like that seen with the F&V stands, the meat stands have developed their own wide network of consumers. By specifically choosing to sell only pork in this particular location, it has developed the label as 'the pork place,' which promotes its status as a semi-formalized space for conducting a particular form of business. Furthermore, New Eisleben Road as a transportation corridor promotes accessibility; people are known to come from all over the Cape Flats region in order to buy the pork that is sold at this particular location.

Spaza shops were found among the housing units in the area surrounding the shopping center. Found in very high densities, their concerns gravitated mainly around competition with other Spaza shops, while competition with the supermarket did not seem to be an issue. In fact, a majority of participants supported Shoprite, as they recognized their own limitations in product diversity, and viewed their Spaza shops not as an alternative to the Shoprite, but providing a different function altogether. When asked about competition with Shoprite, one Spaza shop owner stated,

> No, that's no problem for me, sometimes I don't have something, I only have small stuff. These people need to go to Shoprite … [but] sometimes [Shoprite] don't sell half bread, or something like that. Buy sometimes the half bread or something like that here. (Interview with respondent # 15, 6/27/2012)

While supermarkets were often utilized for big bulk spending at the end of the month (for salaried workers at the higher end of the spectrum, as well as pension and social grant recipients), Spaza shops were seen to provide smaller basic staples on a daily or weekly basis and for those without a consistent income stream. This can be framed as niche construction, and reflect the continued necessity of alternative retail formats beyond the standard formal supermarket. Additionally, it reflects the diverse selection of consumption strategies utilized by consumers. Although some consumers shopped at least occasionally at Shoprite, it was found that many utilized Spaza shops for shorter time frequencies.

In order to employ price differentials as a potential determinant of consumption patterns, a price comparison analysis was conducted between the Spaza shops and Shoprite. Products found at Spaza shops had considerably limited diversity and variety in comparison to the Shoprite. Furthermore, variations in both prices and available product quantities occurred among Spazas, with store size seeming to be a particularly important determining feature. While Shoprite had large bulk options at 5 and 10 kg for basic staples, such as flour, sugar, maize meal, rice, and beans, the majority of Spaza shops interviewed only held options ranging from 500 g to 2 kg, with the occasional 2.5 kg option. For instance, while 9 of 11 Spaza shops stocked 2.5 kg maize meal, only 2 of these stocked the 5 kg option. Conversely, the Spazas held low quantity options unavailable at the Shoprite. For instance, while Spaza shops regularly sold 375 mL bottles of cooking oil, tea containing 10 tea bags, single eggs, and 250 g packets of sugar, these products were unavailable in the supermarket (see Figure 4).

While prices did vary among Spaza shops and price differentials varied between products, it was found that for the majority of products sold at both Spaza shops and the Shoprite, Spaza shops provided them at cheaper prices. There were only few exceptions, such as a can of coke, 2.5 kg packets of maize meal, and 500 g packets of rice. Furthermore, there were certain lower quantity products, such as 375 mL containers of oil, boxes of 10 tea bags and single eggs, which were sold at Spaza shops but missing from Shoprite. These findings reflect the ability of Spaza shops to capitalize on the limitations of the formal market, and develop competitive prices in comparison to their larger, more formalized counterparts. Additionally, it reflects the integration of Spaza shops into formal supply chains, with wholesalers marketing products specifically to these informal sellers. This shows that it is falsely simplistic to claim that supermarkets are pushing Spazas out of business through their use of low prices and economies of scale.[6]

Overall, this research suggests that claiming that consumption patterns reflect a linear path toward greater use of formal outlets is a gross oversimplification. Although consumers do appear to patronize supermarkets more than informal shops, the urban poor in Cape Town utilize the informal sector to compensate for insufficient income to fully participate in the formal economy. Furthermore, Spaza shops are found to be

Figure 4. Price comparison between Shoprite and Spaza shops.
Notes: Graph depicting the difference in prices between the Shoprite supermarket and Spaza Shop products in South African Rands.
Data from primary research conducted by first author (Stephen Peyton).

preferable for day to day purchases, due to their lower prices and smaller quantities, as well as spatial distribution near residential units. The existence of a diverse selection of options supports consumers who value access to the formal supermarket, but are unable to operate only in the formal sector. Therefore, consumers utilize a multiplicity of retail types, both formal and informal, and change consumption patterns to reflect patterns of employment and other means of generating an income.

In townships where supermarkets are present, informal retailers are beginning to construct solid niches alongside formal alternatives. For instance, Spaza shops, although having the same customers and selling the same products as supermarkets, have been able to construct a niche for themselves through supplying slightly cheaper, lower quantity goods. In other words, they capitalize on the limitations of a formal economy in addressing the needs of the poor. It appears as though Spaza shop owners did not see the supermarket as a threat; rather, they accepted this 'competitor' as a vital part of household food procurement strategies. Additionally, F&V stands employ convenience as their main tool in niche construction, while deliberately using the centrality of the supermarket as a way to capture customers. Finally, meat stands position somewhere between the informal and formal, with the social organization and infrastructure of informal business and the supply chains, large consumer base, and wide networks of the formal economy.

On the other hand, supermarkets are limited in their ability to address the needs of the urban poor. While those households with a consistent monthly income would regularly buy from supermarket in bulk, this option is not as readily possible for others working within an informal economy. Within the informal environment of Philippi, wealth disparities and irregular incomes prevent the poorest households from utilizing formal retail outlets. Therefore, the problem is not just limited supermarket expansion, but that the structure of formal retail itself is a restriction on household participation.

The integration of formal suppliers in the supply chains of informal retailers shows how formal practices are being integrated into localized food systems and informal store practices. The supermarket's presence is influencing a renegotiation of retail form, resulting in a hybridized landscape where retailers are reconfiguring their business practices. All Spaza shop owners and workers and F&V stand workers interviewed stated that their products were derived from wholesalers. Moreover, meat stand workers bought their products from a butcher, again capitalizing on the benefits of formalized supply chains. These shifting patterns are mirrored by a greater recognition by wholesalers of the market significance of the informal economy. Wholesalers sell products specially tailored toward informal stores, including smaller quantities of goods such as cooking oil, sugar, and maize meal for Spaza shops (see Figure 4).

This study shows that the traditional view of a linear trajectory toward a fully formalized retail sector is an oversimplification that ignores the capacity and agency of participants in localized food systems. Although it is impossible to rule out this possibility for the future, the flexible business approaches of informal store owners, the complex strategies employed by consumers, and the rapid rate of urbanization all speak to the continued necessity of the structures in place for small-scale, local, informal businesses.

6. Conclusion

Rapid urbanization and high levels of poverty make food insecurity an increasingly problematic phenomenon in Cape Town. Primarily restricted to cash-income for food options with minimal safety nets for resilience, many poor households find that the current infrastructural capacity of the city and public support systems are insufficient to meet their needs.

The expansion of supermarkets in South Africa in the mid-1990s marked a shift toward economic formalization. Food desert literature of the global North views supermarkets as beneficial for food insecure neighborhoods as they provide a wide range of healthy foods priced within a more affordable range than alternatives. This market penetration, however, has been critiqued as a linear economic transformation from informal to formal at the detriment of traditional economies. Under such a scenario, low-income consumers and small shop-owners are excluded from the newly transformed formal economies. Since low-income households have variable consumption patterns contingent on erratic income streams, disease, and social obligations, the format of formal retailers fails to be fully compatible. With formalization exacerbating incompatibility, these communities face a severe loss of resiliency which promotes a higher prevalence of food insecurity.

This article reveals the unequal distribution of supermarkets in Cape Town, with high-income sub-places more likely to have a higher density of supermarkets. Low-income sub-places face drastically lower supermarket densities, indicating reduced accessibility to this food retail format. Supermarket expansion in its current state is constrained by market-based limitations, with a lack of sufficient economic incentive and compatibility in the food systems of the lowest income areas in the city.

The unequal distribution of supermarkets results in a high prevalence of informal food retail formats in low-income neighborhoods in Cape Town. A qualitative case study illuminated the interactions between formal and informal retailers in township environments, revealing the role of supermarkets in household consumption patterns. Research showed that the presence of supermarkets in informal township areas provides alternative methods to obtain food. They also work as anchors for growth and development, which is capitalized upon by informal stores. However, the continued necessity of these informal stores reveals the limitations of the supermarket format in catering to the resilience strategies of the poorest households.

The consumption strategies of the urban poor in Philippi highlight the current incompatibility of poor household consumption strategies with the formalized structure of the supermarket. Many poor households utilize supermarkets for particular spending practices at specific times with limitations for poorest households lacking a consistent income stream. Therefore, while supermarket expansion is significant for improving access to diverse food sources, an automatic assumption of their presence improving food security oversimplifies the relationship between poverty, household resilience strategies, and formal and informal food systems. Furthermore, the dichotomization of informal and formal economies in both literature and policy masks the complex environments in which food access is negotiated. In Philippi, neither a truly formal nor a truly informal economy exists; the economic environment has become a hybridized landscape in which food access is negotiated by the multiple retailers and consumers present. Therefore, a conceptualization of economic activities as distinctly formal or informal falsely represents the reality of these economies. A new way of conceptualizing this retail landscape, and the ways in which consumers negotiate their roles within it, must move away from a sole reliance on large formal retail as the only means of ensuring food security.

In its current state, supermarketization has had some success, although supermarket expansion has been limited in the poorest neighborhoods of the city. Fundamentally, the supermarket has been incapable of fully meeting the needs of the poorest of the poor, who often resort to consumption strategies utilizing a combination of informal and formal food systems. Therefore, a new conceptualization of the interplay of formal and informal retail environments must be accompanied by a more context-specific notion of food deserts to accurately understand food security in Cape Town's townships.

Acknowledgements

We would like to thank Bongi Makanya for her support with field research and as a translator for the first author, Nicholas Lindenberg and Thomas Slingsby from the University of Cape Town GIS Laboratory for their technical assistance in GIS data collection, geocoding, and formatting, Laura Smith from Macalester College for her assistance in statistical analysis, and Paul A. Anderson of the Macalester College Board of Trustee's for providing funding for this project.

Notes

1. Data from the most recent 2011 census was not available at the time the research was conducted.
2. Interviews were divided as such: 11 interviews with Spaza shops, 6 with F&V stands, and 3 with meat stands.
3. 1Rand = .10 USD and .078 Euro.
4. Higher income neighborhoods tend to have lower density settlement and higher access to private transportation.
5. Informal dwellings typically take the form of small residential units assembled out of scrap metal by residents, often referred to as shacks.
6. It must be recognized that these are the views of those shop owners that have adapted to the supermarket's introduction and not representative of business that may have not survived the transition.

Notes on contributors

Stephen Peyton was previously an undergraduate student at Macalester College, currently, pursuing a Masters at the University of Minnesota. He focuses on geospatial analysis of urban social phenomenon, with an emphasis on dynamics of accessibility, food security, and social policy.

William Moseley is professor and chair of Geography, and director of African Studies, at Macalester College. His research interests include political ecology, tropical agriculture, development policy, and food security. Most of his fieldwork has been undertaken in West and Southern Africa.

Jane Battersby is a senior researcher at the African Centre for Cities. Her research focuses on urban food systems and food security within the African context. Her recent publications have focused on the changing nature of food retail in southern African cities, the framing of food security within academia and policy in the African context, and right to food.

References

Abrahams, C. (2010). Transforming the region: Supermarkets and the local food economy. *African Affairs, 109*, 115–134.

Battersby, J. (2011a). The state of urban food insecurity in Cape Town. Urban Food Security Series No. 11. African Food Security Urban Network (AFSUN).

Battersby, J. (2011b). Urban food insecurity in Cape Town, South Africa: An alternative approach to food access. *Development Southern Africa, 28*, 545–561.

Battersby, J. (2012). Beyond the food desert: Finding ways to speak about urban food security in South Africa. *Geografiska Annaler, Series B: Human Geography, 94*, 141–159.

Block, D., & Kouba, J. (2006). A comparison of the availability and affordability of a market basket in two communities in the Chicago area. *Public Health Nutrition, 9*, 837–845.

Crush, J., & Frayne, B. (2011). Supermarket expansion and the informal food economy in Southern African cities: Implications for urban food security. *Journal of Southern African Studies, 37*, 781–807.

D'Haese, M., & Van Huylenbroeck, G. (2005). The rise of supermarkets and changing expenditure patterns of poor rural households case study in the Transkei area, South Africa. *Food Policy, 30*, 97–113.

Department of Finance. (1996). *Growth, employment and redistribution a macroeconomic strategy*. Republic of South Africa, Private Bag X115, Pretoria 0001. Retrieved from http://www.treasury.gov.za/publications/other/gear/chapters.pdf

Food and Agriculture Organization. (1996). *Report of the World Food Summit*. Retrieved from http://www.fao.org/docrep/003/w3548e/w3548e00.htm

Hay, I. (2010). *Qualitative research methods in human geography*. Oxford: Oxford University Press.

Humphrey, J. (2007). The supermarket revolution in developing countries: Tidal wave or tough competitive struggle? *Journal of Economic Geography, 7*, 433–450.

Laderchi, C. R., Saith, R., & Stewart, F. (2003). *Does it matter that we don't agree on a definition of poverty? A comparison of four approaches, Queen Elizabeth House* (Working Paper Series 107). Oxford: University of Oxford.

Larson, J., & Moseley, W. G. (2012). Reaching the limits: A geographic approach for understanding food insecurity and household hunger mitigation strategies in Minneapolis-Saint Paul, USA. *GeoJournal, 77*(1), 1–12.

Laurence, J. (1997). *More equality – Just what the doctor ordered*, The Independent 11 June [online edition]. Retrieved from http://www.independent.co.uk/news/uk/more-equality–just-whatthe-doctor-ordered-1255340.html

Leete, L., Bania, N., & Sparks-Ibanga, A. (2011). Congruence and coverage: Alternative approaches to identifying urban food deserts and food hinterlands. *Journal of Planning Education and Research, 32*, 204–218.

Marshall, M. N. (1996). Sampling for qualitative research. *Family Practice, 13*, 522–526.

Maxwell, D. (1999). The political economy of urban food security in Sub-Saharan Africa. *World Development, 27*, 1939–1953.

Moore, L. V., & Diez, R. A. V. (2006). Associations of neighborhood characteristics with the location and type of food stores. *American Journal of Public Health, 96*, 325–331.

Mubangizi, B. C. (2008). Responses to poverty in post-apartheid South Africa: Some reflections. *International Journal of Social Welfare, 17*, 174–181.

Reardon, T., & Berdegué, J. A. (2002). The rapid rise of supermarkets in Latin America: Challenges and opportunities for development. *Development Policy Review, 20*, 371–388.

Reardon, T., Henson, S., & Berdegue, J. (2007). 'Proactive fast-tracking' diffusion of supermarkets in developing countries: Implications for market institutions and trade. *Journal of Economic Geography, 7*, 399–431.

Reardon, T., & Hopkins, R. (2006). The supermarket revolution in developing countries: Policies to address emerging tensions among supermarkets, suppliers and traditional retailers. *European Journal of Development Research, 18*, 522–545.

Sen, A. (1982). The food problem: Theory and policy. *Third World Quarterly, 4*, 447–459.

Strydom, J. W. (2011). Retailing in disadvantaged communities: The outshopping phenomenon revisited. *Journal of Contemporary Management, 8*, 150–172.

Todes, A., Kok, P., Wentzel, M., Zyl, J., & Cross, C. (2010). Contemporary South African urbanization dynamics. *Urban Forum, 21*, 331–348.

Tschirley, D. (2007). *Supermarkets and beyond: Literature review on farmer to market linkages in Sub-saharan Africa and Asia*. Paper prepared for the AgInfo Project funded by the Bill and Melinda Gates Foundation.

Tustin, D. H., & Strydom, J. W. (2006). The potential impact of formal retail chains' expansion strategies on retail township development in South Africa: Research article. *Southern African Business Review, 10*, 48–66.

ver Ploeg, M., Breneman, V., Farrigan, T., Hamrick, K., Hopkins, D., Kaufman, P., … USA, United States Department of Agriculture. (2009). *Access to affordable and nutritious food: Measuring and understanding food deserts and their consequences: Report to Congress*. Washington, DC: U.S. Department of Agriculture, Economic Research Service.

Weatherspoon, D. D., & Reardon, T. (2003). The rise of supermarkets in Africa: Implications for agrifood systems and the rural poor. *Development Policy Review, 21*, 333–355.
Whelan, A., Wrigley, N., Warm, D., & Cannings, E. (2002). Life in a 'Food desert'. *Urban Studies, 39*, 2083–2100.
Wrigley, N. (2002). 'Food deserts' in British cities: Policy context and research priorities. *Urban Studies, 39*, 2029–2040.
Wrigley, N., Guy, C., & Lowe, M. (2002). Urban regeneration, social inclusion and large store development: The seacroft development in context. *Urban Studies, 39*, 2101–2114.
Zenk, S. N., Schulz, A. J., Israel, B. A., James, S. A., Bao, S., & Wilson, M. L. (2005). Neighborhood racial composition, neighborhood poverty, and the spatial accessibility of supermarkets in metropolitan Detroit. *American Journal of Public Health, 95*, 660–667.

Should I stay or should I go? Incorporating a commitment to fieldwork throughout an academic career

Kristal Jones[a], Matthew A. Schnurr[b], Edward R. Carr[c] and William G. Moseley[d]

[a]*Office of International Programs, College of Agricultural Sciences, The Pennsylvania State University, University Park, PA, USA;* [b]*Department of International Development Studies, Dalhousie University, Halifax, Nova Scotia, Canada;* [c]*Department of Geography, University of South Carolina, Callcott, Columbia, SC, USA;* [d]*Department of Geography, Macalester College, Carnegie Hall, MN, USA*

In this paper, four researchers who share a commitment to applied research and fieldwork methodologies reflect on the ambiguities associated with maintaining and adapting this commitment to changing professional, personal, and contextual situations. The authors focus on the use of fieldwork for the study and support of agricultural change in sub-Saharan Africa, as an example of a setting and topic in which long-term work in the field can improve understanding and support contextualized development. In analyzing a range of experiences associated with maintaining and adapting fieldwork approaches, we complicate and build upon the assertion that professional development pulls international development practitioners and applied researchers away from the field. The experiences analyzed in this paper suggest that the situation of changing orientations toward the field is not dichotomous, and that instead, a commitment to fieldwork can result in innovative approaches to remaining at least partially focused 'outward' and 'downward.' We argue that the epistemological underpinning of situated fieldwork, which recognizes partiality in knowledge and understanding, also requires reflexivity on the part of applied researchers. The reflections and analysis presented here broaden and ground conversations about research ethics, methodological consistencies, and innovative approaches to fieldwork.

Introduction

This paper offers reflections on and analysis of experiences with the opportunities and challenges that come from a consistent methodological and epistemological commitment to fieldwork throughout an academic career. The four authors writing here share common interests in geographical research on agriculture and food systems in sub-Saharan Africa. We have all conducted long-term fieldwork in specific communities and countries on the continent, and have built much of our personal and professional identities on generating place-based knowledge of agrarian livelihoods and change through both qualitative and mixed methods approaches. And, as we discovered in informal conversation at a recent academic conference, we all share a deep awareness of the ambiguities and ethical challenges associated with working in sub-Saharan Africa in particular, and the Global South more generally, as academic researchers based at North American universities with changing professional demands and responsibilities.

Chambers (1997) writes of the moves 'inward' (to the metropoles and ideologies of the Global North) and 'upward' (into management and generalization, away from the particularities field) that tend to characterize the career trajectory of the development professional or academic; an inexorable and seemingly inevitable March away from fieldwork and knowledge of place. Others (Adams, 1979; Edwards, 1989; Freire, 1970) have similarly chastised development practitioners and researchers for their tendency to drift away from the field as they move up the professional ranks, making them more likely to produce research that ignores power dynamics and the ethical implications of work conceptualized and conducted away from the field. The dominant view among these writers is that career advancement leads one away from the outward and downward orientation: one in which a researcher or practitioner is physically based in the field, in a role and with tasks that prioritize local needs and realities over those of outside organizations or institutions. By taking this opportunity to reflect on some of the opportunities and risks that we have faced as scholars of agricultural change in sub-Saharan Africa, we contribute to a literature that wrestles with the ethics and power dynamics embedded within fieldwork (Chambers, 1997; Fox, 2004; Pearson & Paige, 2012). In addition, we hope to personalize such conversations by reflecting on how a commitment to qualitative research methods and fieldwork can remain constant and dynamic, even amid changing professional and contextual demands. Finally, we challenge Chambers' (1997) unilinear characterization of career development leading away from the field as our experiences are significantly more varied.

With this background in mind, this paper offers reflections on the following questions: can long-term fieldwork challenge the 'upward and inward-looking' development research trajectory highlighted by Chambers (1997)? Furthermore, how can a commitment to contextualized qualitative fieldwork remain foundational to research approaches, even as time spent in the field ebbs and flows due to changing professional and personal demands? Fieldwork and the knowledge generated from field-based research methodologies differs depending on the epistemological underpinnings of the researcher, research project, and institutional setting within which the research is carried out. As researchers engaged with qualitative methods and critical analysis have thoroughly documented, it takes more than simply a stated interest in contextualized information to incorporate a reflexive attention to power dynamics and privilege into applied research, especially when that research is conducted by outside academics working in developing countries (for discussions of gender and research, see Cornwall, 2003; Guijt & Kaul Shah, 1998; of professional privilege and research, Chambers, 1997; Edwards, 1989; Greenwood & Levin, 2007; of race and research, Hall, 2005; Kobayashi, 1994).

Much of the often-cited list of challenges facing the self-reflexive African researcher stems from privileging geographical breadth over depth (see Gokah, 2006 for an example). Staying in one place doesn't eliminate such risks, but it does make them easier to manage. We see challenges emerging around an ongoing commitment to fieldwork as requiring recognition of the partiality and contextual nature of all knowledge generation. By recognizing that each researcher and each research approach holds both strengths and limitations, a conversation about knowledge creation, especially in complex socio-ecological environments like much of sub-Saharan Africa, can move beyond a simple dichotomy of upward/downward or place-based/generalizable.

It is not our contention that fieldwork in sub-Saharan Africa, or fieldwork in relation to agricultural research, is unique from fieldwork in other places or on other topics.

However, there are several reasons why a discussion of field research on agricultural change in sub-Saharan Africa is particularly timely and important. From 2000 to 2008, investment in agricultural research and development in sub-Saharan Africa increased by 20%. The food price spikes triggered in 2008 further intensified this commitment to agriculture as the lynchpin of African development (IFPRI, 2011; Moseley, 2011). In addition to increased interest in agricultural development and food security from bilateral development donors, there has also been a dramatic rise in philanthropic giving toward agricultural R&D in sub-Saharan Africa, due mostly to the Bill and Melinda Gates Foundation's establishment of the Alliance for a Green Revolution in Africa in 2006, as well as the more recent emergence of the US Government's Feed the Future Initiative in 2010 (Herdt, 2010; Moseley, 2012; Toenniessen, Adesina, & DeVries, 2008). The current focus on connecting agricultural innovation to poverty reduction in sub-Saharan Africa through a second Green Revolution emphasizes a 'silver bullet' approach to agricultural development that hinges on a few crucial technologies or techniques, and seeks to scale up and out through technology transfer and a reliance on market-oriented development approaches (Brooks, Leach, Lucas, & Millstone, 2009).

The social science research associated with this push for a new Green Revolution for Africa often employs a similar type of generalizability for measuring poverty indicators and impacts of agricultural development projects by relying on quantitative techniques, survey methodologies, and analysis of large, secondary data-sets. Given the trend toward continent-wide, scalable, generalizable agricultural research and development in sub-Saharan Africa, we believe it is more important than ever to critically reflect on the contributions and strengths of doing fieldwork using grounded, qualitative techniques, and maintaining long-term research relationships in specific communities and regions. Qualitative research and fieldwork provide access to different kinds of information and explanation than do quantitative approaches (Patton, 1990; Udry, 2003). This is especially true in sub-Saharan Africa, where relatively little data exist about events on the ground, and where we have such incomplete understandings of the pressures and motivations that shape agricultural decision-making. Long-term commitments to fieldwork in specific communities and regions also provide crucial breadth of contextual knowledge that allows a researcher to combine data gathering and analysis techniques and move from the identification of patterns and correlations to the explanation of these results, producing more robust and rigorous conclusions.

The following sections begin with a discussion of how engaging in long-term fieldwork can support a consistent outward orientation, one that incorporates alternative narratives and tacit knowledge to shape an understanding of place (Moseley & Laris, 2008). Bringing these data into the academic realm challenges and deepens understanding of, for example, agricultural systems in sub-Saharan Africa, which can lead to more appropriate interventions than those proposed by an acontextual second Green Revolution and can help to identify for whom specific interventions will be relevant and desirable. Building on observations of the ambiguous opportunities offered by engaged academic research, we then highlight the challenges associated with maintaining a commitment to long-term, situated fieldwork alongside changing professional and personal demands. We present examples of theoretical, methodological, and logistical approaches that we have utilized to balance this commitment to sustained, grounded research at various stages of our academic careers and to use the resources afforded us by academic positions to support the critical lessons we have learned on the ground. Finally, we reflect on quick decisions to move inward and upward in response to changes in safety and political stability in the field, and the ambiguities inherent in fieldwork that such a move makes explicit. The reflections are written in the first person by each author, in order to convey the personal nature of the subject matter.

Fieldwork with an outward orientation can continually complicate and expand understandings of the field

In this section, Bill and Matthew describe how the practice and process of fieldwork has allowed them to challenge a paradigm of agricultural development that minimizes or obscures local knowledge. They reflect on how a researcher can learn to include multiple perspectives into their own understanding of place, how those perspectives and priorities can be supported through adaptive approaches to research management and local capacity building, and the particular challenges that can arise from a long-term commitment to agricultural research in a single place.

Bill

There is a lot of farming wisdom in the rural African context that is considered 'everyday' knowledge (Rigg, 2007). In many instances, the researcher is attempting to document and describe this local farming knowledge because it is a critical component of understanding management systems (Richards, 1985). Yet its 'everydayness' means that many local informants hardly find local farming knowledge worth mentioning. Further complicating a researcher's quest to document and describe such knowledge may be a long history of top-down agricultural extension wherein local people's farming knowledge was denigrated or belittled. Better farming, people have often been told, lies in adopting new, modern techniques. In my 27 years of working in Africa, from Mali to South Africa, I have repeatedly seen field agents talk down to farmers as if they knew nothing and that their approach was the major obstacle to be overcome. I, for example, have found agricultural extension agents dismissal of local knowledge to be particularly true in Mali (Moseley, 2001). Such pejorative attitudes may be the product of agricultural training in some contexts, and broader political imperatives in others (Davis, 2008).

My most palpable experience with obfuscation involved a 'New Rice for Africa' (Nerica) rice initiative in Mali, following the 2007–2008 global food crises (Moseley, 2011). Some, if not most, of the motivation for this initiative was political as the then Prime Minister was gearing up to run for president in the spring of 2012 (an election which never happened because of a *coup d'état* in March 2012) (Roy, 2010). While my own field work and grass roots reports suggested that this initiative largely failed, the political establishment described it as a huge success, meaning that those in the Ministry of Agriculture felt obligated to describe it in similarly glowing terms. As such, one had to learn how to read between the lines in interviews and create opportunities for informants to talk more informally and off the record. In other instances (most notably in Malawi in my own experience), I have found agricultural extension agents to be some of the most valuable key informants in terms of their broader scale understanding of local food economies (Earl & Moseley, 1996). The trick is to find those agents who have been working in an area long enough to know it well, and regularly travel between villages, so that they have an up-to-date and grounded understanding of food production and trade in the region.

Another strength of fieldwork is that time spent in the field allows one to identify and connect with groups not generally included in agricultural research, or those that are hard to access. While much of African agriculture is dominated by smallholder farms (of which there are roughly 140 million), a not insignificant portion of crop production and animal husbandry occurs on large commercial farms involving the use of farm workers (who number about six million (USAID, 2004)). The views and attitudes of commercial farm workers are important to understand, yet under-researched because this group is often difficult to access and interview under free and confidential conditions. The first challenge is getting onto commercial farms where the farm owner may have little or no interest in talking to a researcher, let alone allowing such a person to interview his/her workers. I repeatedly encountered such difficulties while interviewing commercial farm workers in the Western Cape of South Africa (Moseley, 2007b). The key for me was to get one farm owner to refer me to another (an approach known as 'snowballing' (Patton, 1990)). The second and even more difficult challenge was to get farm owners to allow me to interview their workers (because it meant time away from the job), and to do so in a private and confidential setting. Some farm owners were clearly concerned that I would ask about labor conditions or stir up trouble for them in some other manner. Most farm owners relented to my request for private and confidential interviews once they determined that I was asking questions about relatively benign subject matter (workers' knowledge of farming practices and agroecology in this case). That said, the fact that farm owners continue to have so much power over their workers means that those exploring more politicized issues (e.g. labor conditions) will typically encounter difficulties exploring these questions.

Matthew

One of the many advantages of narrowing the geographic scope of fieldwork in my own research program is that it has created opportunities to build long-term relationships with research assistants (RAs). Much has been written about the struggle to find capable RAs (see for instance Sanjek, 1993; Turner, 2010). I have been working with the same research coordinator since I started my appointment for four years, and my hope is that we will continue working together for the rest of our careers. She is the most important person in this research project (by far!). She coordinates all the logistics, she manages a team of assistants who support her efforts, and she liaises with me constantly. The most important skill she brings to the project is that of relationship building, both with the farmers and with me. She has an incredible gift for relating to farmers, to making them feel comfortable, at ease and valued. She is equally talented at maintaining our relationship: we communicate constantly over email, Skype, and in person. I trust her judgment completely.

I agree with Turner (2010) that the single greatest disservice done by researchers working with RAs is to regard them as 'ghost workers,' and I worry that many researchers do not go far enough in ensuring that assistants are treated as collaborators, rather than employees. This means ensuring that RAs get value out of the project beyond their salary (something that is quite standard for us as academics). One way I have tried to achieve this is by engaging in frank and open conversations with my own research coordinator about what she wants to gain through her involvement with this project. In response, she has emphasized skill development (quantitative data surveys, familiarity with new software such as SPSS), the acquisition of substantive knowledge (insight into debates around agronomy and GMOs) and co-production of outputs (co-presentations at academic conferences, co-publishing). This last point is particularly salient because she a professional herself: one of the goals she set out at the beginning of this project was to increase her participation academic outputs. As such we have worked hard to make sure that her contributions to this project extend beyond just data collection into analysis, results writing, and dissemination. Staying in one place makes it easier to overcome some of the potential hurdles such as power imbalances, communication, and expectations that color the relationship between the researcher and RAs, in order to create long-term collaborations that offer mutual benefit to both parties.

Research approaches that build on knowledge learned in the field can maintain an adaptive downward focus

In this section, Ed and Matthew reflect on incorporating a commitment to fieldwork and qualitative research methods amid changing professional and personal demands. They highlight the need to identify new and adapted methodological and organizational approaches to create continuity and ongoing relevance in field-based projects when extended time in the field is precluded by other responsibilities.

Ed

I cannot debate the changing access to the field that comes with an evolving life and career. When once I thought that a six-week field season was altogether too short, my current childcare and institutional realities lead me to relish the opportunity to spend three weeks in any one place. At the same time, I know that three weeks is simply not enough to build the sort of engagement and understanding that undergird qualitative explanation of events on the ground. My experience of seven field seasons of work, totaling some 20 months, in two communities in Ghana gave me insights into the ways that we talk about globalization, livelihoods, and development that have fueled my engagement with broad debates in development geography (Carr, 2011). These experiences culminated in a recent effort to rethink the entire idea of a livelihoods approach (Carr, 2013), to better align how we think about livelihoods with my extended experience of those livelihoods. Without that time on the ground, I doubt I would have had the insights that drove me to this rethinking. Recognizing the fact I will not, at least for some time, be able to conduct the same kind of long-duration fieldwork has forced me to think about how to implement this rethought livelihoods approach, so that others can use it while I cannot (Carr, 2014). This is critical, because I have a greater belief now, than at any time in my career, in the need for good qualitative fieldwork if we are to understand African agriculture in a meaningful way.

To meet the challenges of maintaining a downward orientation in ongoing research, I have worked to make my handicap into my ally, engaging new pathways of inquiry that I had the luxury to ignore while I was engaging in extended fieldwork. As my time in the field diminishes, my field seasons now look something like this: pre-fieldwork, I spend several weeks preparing materials that will be used to train fieldworkers in the overall philosophy and methods of qualitative research generally, and my particular approach to livelihoods analysis specifically. Upon arrival at the fieldwork site, I designate a week's worth of time to classroom/seminar discussions of these materials with fieldworkers. In this process, we collaborate to make decisions about what methods to use (focus groups, structured/semi-structured interviews, participant observation, etc.) and why, and what questions to ask. In drawing the field teams into this level of project design, they are better able to grasp the underlying goals of the research, and make contributions to the methodology that might facilitate achieving those goals. Once we have a set of agreed methods and tools (question lists for interviews, participatory activities, etc.), we field-test the methodology, usually in multiple sites for several days to a week at each site. At the end of each field test, the team comes back together and discusses the methodology – what worked, what did not work, how the different members of the field team addressed aspects of the methodology, or tools that did not work – and works together to modify/redesign those aspects of the methodology found wanting. Thus, the last step of the research design is, at the same time, a capacity building exercise that facilitates the deep engagement of the field team with the goals of the research, trains them in how to critically evaluate their own methods and results as they go, and teaches them how to rethink their methods to address the shortcomings they identify along the way.

What is at stake, at least in my work, is threefold. The first is the shape of millions of dollars' worth of development programming, overviewed in the introduction to this paper, and aimed at agriculture in the context of a changing climate, which is nothing less than a battle against the 'inward and upward' tendencies described by Chambers (1997). Alongside the shaping of large amounts of development funding, the stakes associated with the place of qualitative research and data in development also include the long-term well-being of many with and for whom we work. When we do not understand what people are doing and why, we often misinterpret those behaviors. For example, farmers in Ghana's Central Region often plant crops in disaggregated plots, sometimes more than a kilometer apart (Carr, 2011). At first glance, this seems like a tremendously inefficient organization of the landscape that forces farmers to spend a lot of time moving between farm plots each day. Qualitative understandings of agricultural practice in this area, however, demonstrates that this disaggregation is a means of hedging agricultural production against seasonal precipitation that is highly variable in terms of amount and timing. While farmers lose some of their production each year as they hedge against excessive rainfall (by planting on the drier tops of hills) and inadequate precipitation (by planting on the bottoms between hills), they are protected against complete crop failure in seasons that bring extremes. Efforts to reorganize this agricultural system could make it more efficient, but at the same time they would likely make it less resilient unless some form of safety net was introduced to protect farmers from catastrophe in a similar manner. Understanding what development is changing is critical to ensuring that any changes are for the better, and not introducing new challenges.

The result of this pivot in my career is a much more serious focus on capacity-building than I might have taken on if I were still able to spend extended stretches in the field, and recognition that I did not emphasize this enough in my earlier work. This is the third issue at stake in my work, creating truly locally owned, and ideally locally generated, research into agricultural and other practices relevant to development and climate change programs. Our conversations about 'upward and inward' presume that the problems we seek to address will endure across entire careers (a rather dismal outlook, when you think about it) but more importantly that people like me, based in an institution in the Global North, will be *needed* indefinitely. Perhaps the only real end to this conversation is one where capacities improve such that we are no longer needed, where our participation becomes optional. If indeed the challenges we seek to address will endure, then this, to me, is what success will look like when I look back on my career.

Matthew

There are many advantages to conducting research in the same place over many years, and much has been written about the creation and execution of culturally appropriate, participatory methods in agricultural research. As Rocheleau (1994) reminds us, the best way to design such a methodological program is to learn from experience: try something out, make mistakes, discuss these mistakes with informants and RAs, and revise accordingly. My current research investigating farmer attitudes and intentions to adopt Genetically Modified banana in Uganda has undergone more than a dozen different iterations (see Schnurr & Mujabi-Mujuzi, forthcoming, for a detailed explanation): first my research coordinator and I went over the entire thing ensuring the wording was appropriate and accurate, before pilot testing it with over 20 farmers and then tweaking protocols based on farmer feedback and our own assessments. We made lots of mistakes. One concerned ranking exercises, where farmers are given numbers (1–15) and asked to rank the ecological constraints to banana production that are most important to them. I assumed that farmers would rank their preferences in ascending order (e.g. 1 is the most important, 15 is the least important), but we quickly realized that farmers ranked their preferences in descending order (e.g. 15 is the most important, 1 is the least important). Another tricky area has been designing proxies for economic wealth: we began by asking farmers how much they spend per week, assuming farmers would feel more comfortable sharing how much they spend rather than how much they earn. This worked quite well for some informants, but others spent so little per week that they preferred giving a value per month. Again, we adjusted the protocols accordingly. Over the 16 months that this project has been ongoing we have experimented with various other proxies for wealth including livestock owned, household electronics, and land size. The proxies for estimating wealth have changed more than 10 times since this research began.

Like Ed, I too have found that long-term fieldwork and place-specific research has become more challenging to undertake over time. As a graduate student, this model of research was relatively easy: I researched South African agriculture by spending years living in South Africa. One potential disadvantage of continuing on with agricultural research in a single African country over a long period of time relates to professional branding. Robertson (1985) argues that developing a deep regional (within the continent) knowledge helps to ward off the homogenizing effects of moving upward to a continent-wide expertise. After completing my doctoral research in South Africa, however, a number of mentors advised me to embark upon my next research project in a different country, so that I would be able to sell myself as an 'Africanist' on the academic job market. Their logic was as follows: most scholars of Africa appointed to positions at North American universities are one of only a handful of Africanists in the faculty (let alone the department). Teaching and supervision responsibilities will include most of the continent, so it's important to be able to demonstrate geographical reach beyond a single country. While at the time I dismissed such suggestions as a bit far-fetched, I am now five years into my career and, reluctantly, acknowledge their legitimacy. I am teaching and supervising over a geographical range that is well beyond my personal expertise: I regularly examine case studies in class on West Africa, for instance, a part of the continent I have never had the chance to visit. Teaching a wide range of undergraduate and graduate courses forces me to draw on a breadth of geographical case studies and examples that might be easier to present if I had privileged breadth over depth in my own research program.

Changing conditions in the field require the incorporation of new knowledge into approaches to fieldwork

In addition to the challenge of navigating changes in personal and professional realities that affect how researchers engage with fieldwork, exceptional events beyond our control often shape the process and substance of fieldwork. In this section, Kristal describes her experiences with making immediate decisions about leaving work in the field due to political instability, and the self-reflection prompted by quick and difficult decisions to move inward and upward, away from the field.

Kristal

Fieldwork and the deep experiential knowledge that results from it is compelling to me in part because long-term familiarity with a place has allowed me to move beyond my own immediate 'otherness,' both by building relationships that create a social role and by providing a material familiarity that allows me to move beyond practical and bodily concerns (Heasley, 2005; Okely, 2007). One important element of this familiarity is the ability to assess both general and more personally specific risks, and to make research and other decisions accordingly (Gokah, 2006). Awareness of personal safety and relative risk is a fairly constant element of fieldwork in my experience, but decisions based on that awareness are often minor and occur within the overall flow of research and time in the field. The decision to fundamentally alter or call off research based on an unexpected event or new potential threats has proved challenging and at times gut-wrenching for me, in part because of the understanding I thought I had of my own place in the field.

I was conducting fieldwork in Mali in March 2012 when a military coup occurred. I was on the outskirts of Bamako at a research station when all foreign staff received text messages from their respective embassies warning of possible instability and violence. The first round of decision-making, to get to the airport before town shut down in chaos and to buy an on-the-spot seat in the only plane leaving that night, happened so fast there was little time to reflect. At the same time, it felt like an excruciating admittance of many small, often ignored details about a Westerner doing fieldwork in Africa. On the one hand, I wanted to get out if at all possible, given the uncertain nature of what was happening and how it could all play out in coming days and weeks. On the other hand, what did that say of my commitment to Mali and West Africa, that at the first sign of real danger or crisis I wanted to leave? More immediately, the fact that I both wanted and had the ability to pull out a credit card and buy a plane ticket out was a stark reminder of the difference between long-term commitment to work in and care for a place, and actually being from that place.

This divide is of course always present when doing fieldwork – we might spend weeks or months in villages, eat local food, dress appropriately, learn local languages, and even assume identities that approximate localness. However, we can and do drop those identities at some point to return home, and we can choose when and how that happens. From the point of view of some sociological theory (e.g. Berger & Luckmann, 1967), this description of the social construction of identity and reality is a part of all human interaction, and is perhaps not so exceptional. However, in the context of fieldwork in sub-Saharan Africa, the differences between there and here are stark, and, perhaps because they are so material, they are at times unsettling and difficult for me to reconcile. Making the decision to leave Mali in a moment of crisis was an encapsulation of this ongoing unease, which stems from the contrast between a feeling of genuine commitment to being there, and the begrudging acknowledgment that I am different and 'privileged,' and perhaps an interloper in others' realities. For me, a moment of crisis and immediate thoughts about safety brought up underlying ethical concerns I have about committing to long-term relationships through fieldwork, while recognizing that I am also maintaining distance and have the ability to leave or end the relationships whenever I choose.

In the aftermath of the coup in Mali, there have been many rounds of discussion amongst colleagues about how to continue research given new security concerns and uncertainties, if and when research (particularly longitudinal data collection) is less important than personal safety, and how to make those decisions both for ourselves and with our collaborators. This last point is particularly challenging, I think, as everyone will have their own threshold for what they are willing to risk in the name of research. I was planning to spend spring 2013 in Mali and Niger to do another round of data collection for an ongoing project, and after several months of news watching and soliciting all possible opinions, made the decision that I did not feel safe doing fieldwork alone in remote rural areas. This decision was met with varying degrees of agreement (or not) by my colleagues and collaborators, with some expressing almost disgust that I would let 'sensationalistic' media reports and the random, admittedly small possibility of kidnapping stand in the way of research commitments and good data. Working through my own priorities in these decisions included separating out my feelings about commitment to West Africa and the possibility that a place to which I feel connected might have changed in ways that I don't yet understand, a desire to do rigorous (so as to be meaningful) research, and my own feelings about privilege and mobility.

Discussion

The above examples and reflections on the role of fieldwork in maintaining grounded research agendas offer an additional layer of critical analysis and nuance to Chambers' (1997) assertion that professional advancement and changing personal circumstances inevitably push researchers in the realm of international development inward and upward on the scales of ideological and political power. For many of us, and others who have written for this journal, Chambers' positive ideal-type reflects our trajectories in the field: starting out as Peace Corps volunteers (as two of us were) or PhD students conducting dissertation research (as all four of us did), we tend to spend less and less time on fieldwork as our careers and personal lives progress (see also Newhouse, 2012; Pearson & Paige, 2012). But such outcomes are neither linear nor deterministic. We believe it is possible to develop a fieldwork ethic that remains committed to outward and downward research despite these challenges: one based on humility, gratefulness, collaboration, and generosity. Make no mistake: this paper was generated not by back-patting ourselves for maintaining this commitment. Rather, it emerged from recognition that while many complicating and legitimate life factors can challenge our ability to meet Chambers' positive ideal-type, researchers can also approach these changes as opportunities to create new models for agricultural research that maintain a commitment to outward and downward research.

Our experiences related to adapting to the ambiguities of integrating a fieldwork ethic into complex professional and personal demands suggest that there are many pragmatic approaches to fieldwork and place-based research that generate meaningful information and experiences that continue to inform the research processes. We believe that the ideal-types described by Chambers (1997) are useful as bookends to a spectrum of approaches and positions within the field of research for development, but that these ideal-types should not be reified to the exclusion of innovative ways of resisting such pulls via a fieldwork ethic. In Figure 1, we suggest a messier, more realistic typology that depicts the practical combinations of decision-making and approaches to research

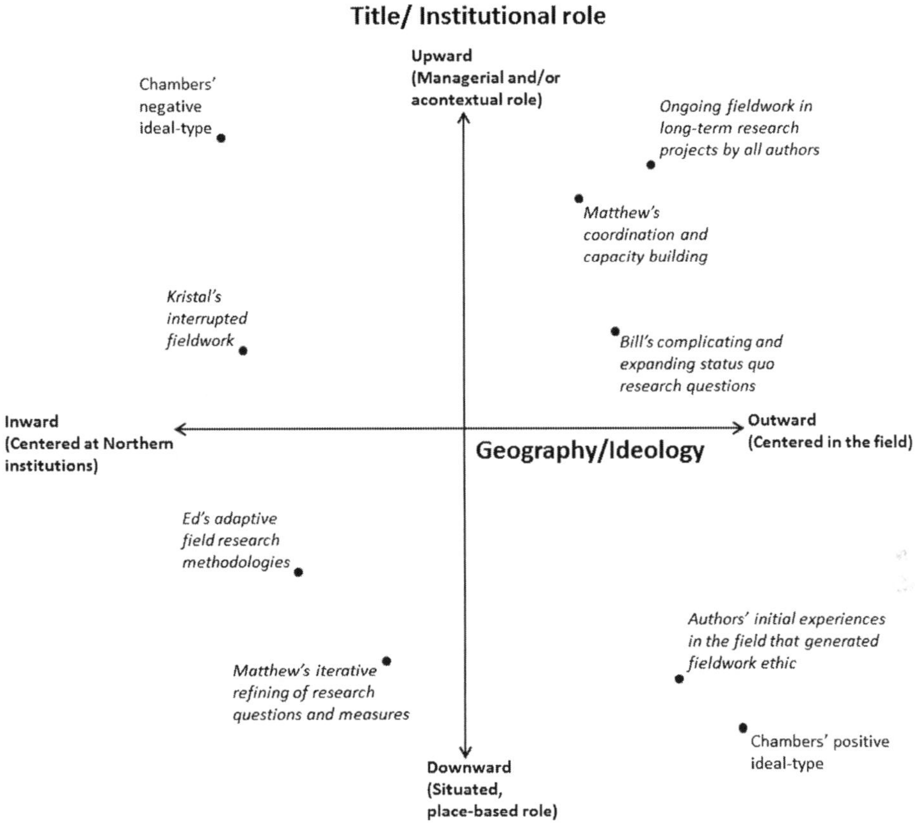

Figure 1. Elements and examples of adaptive fieldwork in agricultural research for development.

for development that are described in this paper. We situate aspects of each author's narrative within a dynamic and temporally adaptive view of Chambers' two spectra of inward-outward and upward-downward.

In our view, ethical fieldwork begins with the conviction that no single person in the research process has all of the answers, or all of the questions – an acknowledgment that is particularly salient within agricultural research in sub-Saharan Africa, where agricultural and social systems are highly diverse and complex (Richards, 1989). Figure 1 demonstrates both the complex nature of research conducted in and about the field, and also the partiality of any single research approach or step of a research project. Bill's description of how fieldwork and expert status allows him to include voices and knowledges not necessarily privileged by either development discourse or local elites complicates the negative connotations associated with moving away from the field due to changes in title or rank. The status and financial means that come with an academic appointment at a North American university can provide the resources needed to ask more nuanced questions in the field than would be otherwise possible. Similarly, Matthew's example of capacity building among local RAs and managers describes an upward shift in his role in the field. But by engaging in a long-term research project and ongoing fieldwork, the research process has remained focused on the needs and realities of the field site, as evident his later description of evolving metrics, methods and analysis.

Conducting research on agriculture, rural livelihoods, and agrarian change in sub-Saharan Africa is different than implementing development projects addressing needs in the same substantive and geographic areas. Most of us have worked as field-based development professionals, and sought out scholarly research as a means of retaining a downward-oriented role in the research-for-development process (Moseley, 2007a). Interestingly, however, this inward shift has created space and time for conducting primary fieldwork and developing adaptive research methodologies and analytical concepts that reflect the on-the-ground realities of rural agricultural communities. Ed's example of conducting primary fieldwork that is place-specific continues to guide his theoretical and academic work. It also suggests that the critical thinking and support for primary research fostered in university contexts can create opportunities for researchers to continue to engage in long-term research that generates outputs that are meaningful for both the local context and the (inward-looking) academic realm. The collaborative, iterative process described by Matthew of exploring and refining research questions and indicators of change reflects a critical and situated epistemological approach to research reflects a continued downward orientation of a researcher's role in the field, despite being conducted largely from afar.

Finally, Kristal's example of fieldwork interrupted challenges Chambers' (1997) insinuation that moving upward and inward is invariably done by volition or with insensitive intent. The ambivalent feelings that accompany all moves into and away from the field are brought into high relief during moments of insecurity and uncertainty. Sometimes circumstances beyond our control force a researcher away from the field. These changing roles can happen in moments of crisis, as Kristal's narrative recounts, or in response to professional and personal circumstances that change more slowly. Either way, the original downward and outward orientation that many researchers appreciated and adopted during initial long stints in the field can persist through careful and committed reflexivity and relationships.

Conclusion

The examples provided in this paper describe how each of us has navigated changing professional, personal, and contextual realities that have pushed us 'away' from the field. We see these challenges as inherent in all fieldwork decisions, including the decision to go in the field in the first place, and to repeatedly return. Self-reflexivity is a crucial counterpart to qualitative research methods (Naples, 2003), and all of the authors here turn a critical lens on their own commitments to fieldwork to better understand the strengths, challenges, and potential contradictions of being a North American researcher engaged in long-term research in Africa. Wrestling with such sticky and sensitive issues allows us to learn from and support each other, and reminds us that we need to be vigilant in working towards a more inclusive and realistic understanding of the fieldwork process, as well as its impacts and limitations. Being explicit about the trade-offs that exist within the decisions we make as researchers creates a more humble and honest assessment of our own struggles to retain a commitment to fieldwork and applied research amidst changing professional and personal circumstances. We offer these personal experiences and reflections in the hope that they might be useful for others as they encounter the challenges associated with research into agricultural development in Africa.

Notes on contributors

Kristal Jones is a postdoctoral researcher in the Office of International Programs, College of Agricultural Sciences at Penn State. She is a rural sociologist with research interests in seed systems, the social impacts of agricultural research for development, and the political ecology of agricultural science. Her work is focused on Francophone West Africa.

Matthew Schnurr is an associate professor in the Department of International Development Studies at Dalhousie University. He is an environmental geographer with research and teaching interests in environment and development, political ecology, agricultural biotechnology, farmer decision-making, and environmental justice. His work focuses primarily on eastern and southern Africa.

Edward R. Carr is an associate professor in the Department of Geography at the University of South Carolina. His interests lie broadly in the challenges of development and global change. He directs the Humanitarian Response and Development Lab (HURDL) at USC, which bridge the worlds of research and implementation in various manifestations of these broad challenges.

William G. Moseley is a full professor and department chair in the Department of Geography at Macalester College. His research interests include political ecology, tropical agriculture, environment and development policy, and livelihood security. Most of his fieldwork has been undertaken in West and Southern Africa.

References

Adams, A. (1979). An open letter to a young research. *African Affairs, 78*, 451–479.

Berger, P., & Luckmann, T. (1967). *The social construction of reality in everyday life*. Garden City, NY: Anchor Books.

Brooks, S., Leach, M., Lucas, H., & Millstone, E. (2009). *Silver bullets, grand challenges and the new philanthropy*. STEPS working paper no. 24. Brighton: STEPS Centre.

Carr, E. R. (2011). *Delivering development: Globalization's shoreline and the road to a sustainable future*. New York, NY: Palgrave Macmillan.

Carr, E. R. (2013). Livelihoods as intimate government: Reframing the logic of livelihoods for development. *Third World Quarterly, 34*, 77–108.

Carr, E. R. (2014). From description to explanation: Using the livelihoods as intimate government (LIG) approach. *Applied Geography, 52*, 110–122.

Chambers, R. (1997). *Whose reality counts? Putting the first last*. London: Intermediate Technology Publications.

Cornwall, A. (2003). Whose voices? Whose choices? Reflections on gender and participatory development. *World Development, 31*, 1325–1342.

Davis, K. (2008). Extension in sub-Saharan Africa: Overview and assessment of past and current models and future prospects. *Journal of International Agricultural and Extension Education, 15*, 15–28.

Earl, J., & Moseley, W. G. (1996). *Risk map food security baseline: Final report for Malawi*. London: Save the Children Fund (UK).

Edwards, M. (1989). The irrelevance of development studies. *Third World Quarterly, 11*, 116–135.

Fox, R. C. (2004). Observations and reflections of a perpetual fieldworker. *The Annals of the American Academy of Political and Social Science, 595*, 309–326.

Freire, P. (1970). *Pedagogy of the Oppressed*. New York, NY: Continuum.

Gokah, T. (2006). The naïve researcher: Doing social research in Africa. *International Journal of Social Research Methodology, 9*, 61–73.

Greenwood, D. J., & Levin, M. (2007). *Introduction to action research*. Thousand Oaks, CA: Sage.

Guijt, I., & Kaul Shah, M. (Eds.). (1998). *The myth of community: Gender issues in participatory development*. London: Intermediate Technology Publications.

Hall, B. (2005). In from the cold? Reflections on participatory research from 1970–2005. *Convergence, 38*, 5–24.

Heasley, L. (2005). On walking contested land: Doing environmental history in West Africa and the United States. *Environmental History, 10*, 510–531.

Herdt, R. W. (2010). People, institutions, and technology: A personal view of the role of foundations in international agricultural research and development 196–2010. *Food Policy, 37*, 179–190.

IFPRI (2011). *2011 global food policy report*. Washington, DC: International Food Policy Research Institute.

Kobayashi, A. (1994). Coloring the field: Gender, "race", and the politics of fieldwork. *The Professional Geographer, 46*, 73–80.

Moseley, W. G. (2001). *Sahelian 'white gold' and rural poverty-environment interactions: The political ecology of cotton production, environmental change, and household food economy in Mali* (PhD dissertation). Department of Geography, University of Georgia, Athens, GA.

Moseley, W. G. (2007a). Collaborating in the field, working for change: Reflecting on partnerships between academics, development organizations and rural communities in Africa. *Singapore Journal of Tropical Geography, 28*, 334–347.

Moseley, W. G. (2007b). Neoliberal agricultural policy versus Agrarian justice: Farm workers and land redistribution in South African's Western Cape Province. *South African Geographical Journal, 89*, 4–13.

Moseley, W. G. (2011). Lessons from the 2008 global food crisis: Agro-food dynamics in Mali. *Development in Practice, 21*, 604–612.

Moseley, W. G. (2012). *The corporate take-over of African food security and agricultural production*. Good Governance Africa, November 1. Retrieved from http://gga.org/stories/editions/aif-6-whose-land-is-it-anyway/the-corporate-take-over-of-african-food-security-and-agricultural-production

Moseley, W. G., & Laris, P. (2008). West African environmental narratives and development-volunteer praxis. *Geographical Review, 98*, 59–81.

Naples, N. (2003). *Feminism and method: Ethnography, discourse analysis, and activist research*. New York, NY: Routledge.

Newhouse, L. S. (2012). Footing it, or why I walk. *African Geographical Review, 31*, 67–71.

Okely, J. (2007). Fieldwork embodied. *The Sociological Review, 55*, 65–79.

Patton, M. (1990). *Qualitative evaluation and research methods*. New York, NY: Sage.

Pearson, A. L., & Paige, S. B. (2012). Experiences and ethics of mindful reciprocity while conducting research in Sub-Saharan Africa. *African Geographical Review, 31*, 72–75.

Richards, P. (1985). *Indigenous agricultural revolution: Food and ecology in West Africa*. London: Hutchinson.

Richards, P. (1989). Agriculture as performance. In R. Chambers, A. Pacey, & L. A. Thrupp (Eds.), *Farmer first: Farmer innovation and agricultural research* (pp. 39–43). London: Intermediate Technology Publications.

Rigg, J. (2007). *An everyday geography of the global south*. New York, NY: Routledge.

Robertson, A. F. (1985). Doing research in Africa. *African Affairs, 84*, 279–290.

Rocheleau, D. (1994). Participatory research and the race to save the planet: Questions, critique and lessons from the field. *Agriculture and Human Values, 11*, 4–25.

Roy, A. (2010). L'initiative rix au Mali: Une réponse politique à l'insécurité alimentaire [The rice initiative in Mali: A political response to food insecurity]. *Politique Africaine, 119*, 87–106.

Sanjek, R. (1993). Anthropology's hidden colonialism: Assistants and their ethnographers. *Anthropology Today, 9*, 13–18.

Schnurr, M. A., & Mujabi-Mujuzi, S. (forthcoming). 'No one asks for a meal they've never eaten'; Or, do African farmers want Genetically Modified crops? *Agriculture and Human Values*.

Toenniessen, G., Adesina, A., & DeVries, J. (2008). Building an alliance for a green revolution in Africa. *Annals of the New York Academy of Sciences, 1136*, 233–242.

Turner, S. (2010). Research note: The silenced assistant. Reflections of invisible interpreters and research assistants. *Asia Pacific Viewpoint, 51*, 206–219.

Udry, C. (2003). Fieldwork, economic theory and research on institutions in developing countries. *American Economic Review, 93*, 107–111.

USAID. (2004). *Poverty reduction and agricultural trade in Sub-Saharan Africa: recommendations for USAIF interventions*. Discussion paper. Washington, DC: US Agency for International Development.

Food and nutrition assistance to HIV-infected and affected populations in Ghana: a situational analysis and stakeholder views

Amos Laar[a], Angela El-Adas[b], Richard N. Amenyah[b], Kyeremeh Atuahene[b] Elizabeth Asare[c], Eric Y. Tenkorang[d], Matilda Laar[e], Andrew Anthony Adjei[f] and Isabella Quakyi[g]

[a]Department of Population, Family, & Reproductive Health, School of Public Health, University of Ghana, Accra, Ghana; [b]Ghana AIDS Commission, Cantonments, Accra, Ghana; [c]Nestle Ghana Limited, Accra, Ghana; [d]Department of Sociology, Memorial University, St. John's, Canada; [e]School of Dietetics and Human Nutrition, McGill University, MacDonald Campus, Bellevue, Montreal, Canada; [f]Department of Pathology, Korle Teaching Hospital, College of Health Sciences, University of Ghana, Accra, Ghana; [g]Department of Biological, Environmental, and Occupational Sciences, School of Public Health, University of Ghana, Accra, Ghana

Poor nutrition is a determinant of death particularly among persons living with HIV (PLHIV). This paper discusses the state of food and nutrition assistance to PLHIV in Ghana. A review of published articles and program reports was done and supported by in-depth interviews with key stakeholders. Data were qualitatively analyzed. A matrix highlighting problems that need to be addressed, possible attributable factors, strengths, and opportunities of existing actions metamorphosed from the analysis. Although stakeholders agree that adequate nutrition is critically important for PLHIV, food and nutrition interventions to PLHIV are inadequate.

Introduction

A progress report published by WHO, UNAIDS, and UNICEF in 2011 shows that the treatment component of the response to the HIV epidemic in resource-poor settings is beginning to receive the attention that it deserves. The coverage of antiretroviral therapy among persons living with HIV (PLHIV) in low- and middle-income countries increased from 400,000 people in 2003 to 5.25 million by the end of 2009 (WHO, 2011). Sub-Saharan Africa had the highest increase in the number of people receiving antiretroviral treatment (ART) in 2010, from 3,911,000 in December 2009 to about 5,064,000, representing a 30% increase in one year. Unfortunately, this improvement translates into providing treatment to only 49% of people in need of ART. The introduction of ART, according to the report, has averted 2.5 million deaths in low- and middle-income countries globally since 1995, but mortality among PLHIV remains unacceptably high in developing countries (WHO, 2011).

It has been documented that the many factors contributing to the high levels of mortality in PLHIV include chronic malnutrition (caused by inadequate consumption of food and essential nutrients due to food insecurity and loss of appetite) and also from infection or disease (Houtzager, 2009; Villamor et al., 2006). There is currently a plethora of scholarship on the interactions of malnutrition and HIV. Malnutrition leads to immune impairment, worsens the effects of HIV, and contributes to a more rapid progression of the disease (Gillespie & Kadiyala, 2005). Even though treatment with antiretroviral drugs is effective in inhibiting viral replication and decreasing viral load, nutrition is directly relevant to treatment as interactions between these drugs and nutritious food can significantly influence the success of ART by affecting drug efficacy and adherence to drug regimes. In this regard, nutrition is as important as access to antiretroviral therapy in every setting. Adequacy of nutrition is particularly critical to maximizing the benefit of ART to PLHIV in resource-poor settings.

Existing evidence indicates that HIV and food/nutrition insecurity are connected in a vicious cycle. While food insecurity increases susceptibility to HIV exposure and infection, HIV in turn increases vulnerability to food insecurity (Loevinsohn, 2009). Thus, food insecurity may place people in situations of heightened risk of exposure to the virus. For instance, people who are food insecure may be less able to access information about HIV or less able to act on their knowledge of risk to minimize HIV exposure. They may also be forced to separate from their families to earn a livelihood, and in the case of women, into transactional sex in order to feed their families (Gillespie & Kadiyala, 2005).

The manifestation of the global resolve to address this problem is the recent emphasis by several international agencies that food and nutrition interventions should be systematically linked to antiretroviral interventions. The widespread overlap of HIV and malnutrition in sub-Saharan Africa requires that clinical and public health interventions be complemented by nutrition and food assistance programs (Rawat, Kadiyala, & McNamara, 2010). The World Bank, for example, called for rapid action research and learning by doing in order to scale-up the integration of nutrition programs in the context of HIV and AIDS (The World Bank, 2006). A review of approaches to food and nutrition in HIV programs in Africa concluded that current HIV and AIDS policies have tended toward highly medicalized approaches and called for a comprehensive approach to link health strategies with community-oriented food-based strategies (Panagides, Graciano, Atekyereza, Gerberg, & Chopra, 2006).

Shannon and colleagues have called for structural HIV interventions that incorporate food assistance strategies for PLHIV (Shannon et al., 2011). In addition, a study stated that food support programs may be required in addition to nutrition support to optimize nutritional status and health outcomes in PLHIV who are food insecure (Houtzager, 2009). Another study in British Columbia found that food insecurity is a risk factor for mortality among antiretroviral therapy-treated individuals, particularly among individuals who are underweight, and calls for innovative approaches to address food insecurity to be incorporated into HIV treatment programs (Weiser et al., 2009). Modest nutritional education and supplementation have been found to be associated with improved nutritional status and quality of life of PLHIV (William, 2009).

In settings such as Ghana, nutrition interventions including food assistance is relevant not only to PLHIV but also to the general population. In 2009, the World Food Program (WFP) assessed food insecurity in Ghana using an indicator that combined household food consumption frequency and dietary diversity and reported that about 5% of Ghanaians (about 1.2 million) were food insecure (WFP, 2009). An additional 9%

(about 2 million) were at risk of food insecurity. The problem is even worse among HIV-affected households, according to a recent national survey specifically for HIV-affected populations. The survey reported that 16% of the HIV-affected Ghanaian households were food insecure (WFP & GAC, 2011).

Aware of the weaknesses in the local nutrition programs and services, the Food and Nutrition Technical Assistance Project (FANTA) collaborated with the Ghana Ministry of Health (MOH), the Ghana Health Service (GHS), and other nutrition stakeholders to strengthen nutrition programs and services and integrate them into the Ghanaian health system. From 2007 to 2013, these efforts helped introduce community-based management of acute malnutrition (CMAM) and scaled up integrated CMAM services within the existing Ghanaian health system. They also introduced nutrition assessment, counseling, and support (NACS) and scaled up integrated NACS services within existing HIV and tuberculosis (TB) service delivery, and have strengthened maternal and child health and nutrition (MCHN) services through technical support on advocacy and coordination (FHI 360 & FANTA, 2014).

This paper presents a situational analysis of food and nutrition assistance to PLHIV and affected populations in Ghana. It documents problems in existing programs needing attention and proposes actions to address them.

Methods

A literature review (including peer-reviewed articles, compiled program documents, reports, and anecdotes) was carried out. We also conducted in-depth interviews with 17 key stakeholders involved at various levels of Ghana's response to HIV and AIDS. These included government agencies such as the Ghana AIDS Commission (GAC), the National AIDS/STI Control Program (NACP), HIV treatment centers, multilateral agencies, bilateral agencies, local and international nongovernmental organizations, and the national association of positive persons (NAP+). Three of the 10 regions of Ghana were purposively selected for this exercise – northern region, eastern region, and the Greater Accra region. We summarize below the reasons that informed the selection of these regions.

Even though Ghana as a country is categorized as a lower middle income food deficit country, the situation is much worse in parts of northern Ghana where up to 9 out of 10 people live below the poverty line (Ghana Statistical Service, 2008). Data from a nationwide survey also show that populations located in the northern region are the most food insecure people in the country (WFP, 2009). Although HIV prevalence in northern Ghana has been the lowest over years (NACP, 2013), the region is prone to extreme weather conditions. This region suffers from periodic floods and droughts, the worst of which in recent history occurred in 2007. The eastern region on the other hand has one of the highest HIV prevalence in Ghana with HIV prevalence in one of the sentinel sites in the region ranging from 18% in 1992, 5.8% in 2009 (NACP, 2011), and 3.6% in 2012 (NACP, 2013). The HIV prevalence for the Greater Accra region carries a mixed picture. From a high prevalence of 4.3% in 2003, there was a reduction to 2.1% in 2005. However, since 2006 there has been a gradual increase in the HIV prevalence in the Greater Accra region. According to the 2009 HIV sentinel report, the HIV prevalence for the region was 3.2% (NACP, 2011), which is almost double the national HIV prevalence. The latest HIV prevalence in the region is 3.5% compared to a national average of 1.3 (NACP, 2013). Also as the national capital, the region is home to most of the stakeholders needed for this analysis.

Visits were made to such stakeholders as the Nutrition Department of the GHS, the NACP, the GAC, UN agencies (UNAIDS, WHO, UNFPA, UNICEF), multilateral agencies (the World Bank and the Global Fund), bilateral agencies (USAID, FANTA, DANIDA, CIDA), and international and local NGOs (OICI, ADRA, CRS). The NAP+, staff of 4 treatment centers and 14 PLHIV (two each from OICI, CRS, Tema General Hospital Treatment Centre, Korle-Bu Teaching Hospital Fevers Unit, St. Martins de Porres Hospital Treatment Centre, Atua Government Hospital Treatment Centre, the Tamale Central Hospital) who had benefited from food assistance interventions were also interviewed. A faculty member from the Department of Nutrition and Food Science of the University of Ghana, the International Labour Organization, and the Ghana Employers Association also participated in the interviews.

With the help of an in-depth interview guide, the data that were gathered from these stakeholders included their current activities on HIV and AIDS, their involvement in food and nutrition support interventions, and their ranking of nutrition and food support among other priorities. Their knowledge and use of guidelines on HIV and AIDS and nutrition, policy documents on food and nutrition support to PLHIV, standards of practice on the nutritional management of PLHIV was also documented. Partnerships with other stakeholders and perceived gaps in food and nutrition support interventions in Ghana were equally documented. We interacted with and documented concerns from the executive members of NAP+, PLHIV, families/caregivers, and service providers.

The data collected were analysed manually using qualitative procedures. This produced transcripts that were then sorted into thematic areas, from which gap analysis matrix metamorphosed. First, the handwritten field notes were analysed manually using the principles of systematic text condensation as described by Malterud (1993). It involved four steps – repeated review of the notes to gain a thorough sense of the overall content in the texts, identifying central meaningful units in the material, condensation of the content through a coding of the text, and finally creating categories that contained the condensed meaning of the main themes in the material. The outputs were then sorted into thematic areas. Attempts were made to corroborate the views derived from the interviews with those captured in the various reports.

Results

The National HIV Sentinel Surveillance report of Ghana indicates the total number of HIV-infected adults to be 225,478 and that of children to be 30,439 (NACP, 2011). Based on an average household size of four, over 900,000 persons are directly affected by HIV. A national survey commissioned by the WFP and GAC found that 16% of 1745 HIV-affected households who participated in the survey were food insecure. The households sampled were those whose members were registered for ART. The survey further estimated that conditional household support/social protection interventions are needed for some 40% of all PLHIV on ART (WFP & GAC, 2011).

A desk review by Owusu-Amoako (2013) on the integration of nutrition indicators into Ghana's HIV and tuberculosis monitoring and evaluation revealed that nutrition indicators as of 2013 were not collected through the system managed by the NACP. Key stakeholders now agree on portions of the HIV client folders and report forms where nutrition indicators should be included. The national Tuberculosis Control Program has already incorporated NACS indicators into the pediatric and adult client cards and reporting formats at the facility, district, regional, and national levels (Owusu-Amoako, 2013).

Food and nutrition support initiatives by stakeholders in Ghana

The analysis shows that about 6000 PLHIV were covered by the food assistance interventions of the Catholic Relief Services (CRS) and OICI. Records from the NAP+ secretariat reveal that about 30,000 of the over 200,000 PLHIV or the over 900,000 affected population in need of food assistance were covered nationally. We argue based on these that food and nutrition assistance to HIV-infected and affected populations in Ghana are inadequate. Even though stakeholders rate food security and nutrition interventions highly, they have contributed minimally to its implementation. A stakeholder noted that a lot of attention is centered on other non-nutrition-related but equally important activities. These activities range from coordination and management of the overall national HIV response, to providing technical assistance on nutritional assessments and counseling. Summarized in the following paragraphs are the existing efforts by various stakeholders.

Efforts of the Ghana AIDS Commission

The Ghana AIDS Commission was formed in 2000. It was legally established by ACT 613 (2002) of the Parliament of the Republic of Ghana, as a supra-ministerial and multi-sectoral advisory body to direct and coordinate the involvement of all development partners, ministries, the private sector, and civil society organizations including NAP+ in the response against HIV epidemic in Ghana.

As the highest policy-making body on HIV and AIDS, GAC provides effective leadership in the coordination and management of programs and activities undertaken by all stakeholders of the national response through evidence-informed advocacy, joint planning, research, monitoring and evaluation towards comprehensive prevention and total control of the disease within the context of the National Strategic Plan (NSP). Furthermore, the GAC is constantly in the drive to mobilize resources internally and externally for sustainable national HIV response.

Food and nutritional interventions form essential components of the NSP under treatment, care, and support. This has been implemented with technical assistance from the development partners. Several initiatives and strategies have been adopted by partners to deliver food and nutritional interventions to beneficiary communities in the country. The GAC partnered with USAID to implement the Food for Peace (FFP) program, which expired in 2010. GAC has partnered with the US government through the FANTA project among others provides Food by Prescription for PLHIV. The GAC, with funding from DANIDA, is implementing a home-based care program for PLHIVs in the eastern, central, and western regions of Ghana. This program entails, among other things, training on comprehensive institutional and home-based care and support as well as strengthening volunteer systems in the communities. The program is being implemented by OICI with support from the Christian Health Association of Ghana (CHAG), CRS, and other stakeholders. Furthermore, GAC is partnering with the WFP on how to support Ghana's HIV programs in the food insecure areas of the country especially in the three northern regions.

Efforts of the United Nations WFP

The WFP focuses on reducing chronic hunger and malnutrition, strengthening the country's capacity to reduce hunger and restoring and rebuilding lives after emergencies. To achieve this, WFP supports the government's school feeding program, and provides women and children at risk of malnutrition with complementary fortified foods. A related joint WFP/UNICEF project provides women's group with income-generating activities, which reduce malnutrition in their communities. WFP has developed an active local food procurement program and has a protracted recovery and relief operation (PRRO) underway, aimed at improving the food and nutrition security of people who were severely affected by a combination of floods, droughts, high food prices, and the global economic crisis. The PRRO provides supplementary feeding to malnourished children, pregnant and nursing women, and food insecure PLHIV. In partnership with the GAC, the WFP is supporting Ghana's HIV Programs especially in the food insecure areas of the country (the three northern regions).

Efforts of the World Bank and the WHO

Our interaction with the World Bank reveals that their main priority lies in strengthening health systems rather than food assistance. They provide help to strengthen systems, and behavioral change to ensure good/adequate/sustainable nutrition. The World Bank ranks nutrition as medium on a scale of 'high, medium, and low'.

As a technical agency with particular interest in nutrition, the WHO, even though does not implement nutrition interventions at the field/community level, does support the MOH in its implementation of nutrition interventions. The WHO country office suggests that all interventions need to be coordinated by the government and standards of operation made clear to all the partners.

Effort of the FANTA

The FANTA works with the government of Ghana and US President's Emergency Plan for AIDS Relief to help improve HIV care and treatment services through the introduction of nutrition care and support into national policies, strategies, and programs. The project is currently offering technical assistance to GHS and NACP on strengthening the integration of food and nutrition into Ghana's national response to HIV by offering NACS. The project produces and disseminates program guidance on nutritional care and support interventions, the nutrient requirements of PLHIV, and food and nutrition implications of ART. The Ghana HIV and Nutrition Job Aids (2011) is a useful example: these job aids assist clinicians and community health workers working in inpatient and outpatient HIV services to assess nutritional status and provide critical nutrition actions and information to people living with HIV.

FANTA facilitated the provision of ready-to-use therapeutic foods and fortified blended foods for severely and moderately acute malnourished persons. The target groups include PLHIV, pregnant women, lactating women, and children. Phase 1 of this was piloted in 10 clinics and has since been scaled up. The FANTA project collaborated with the Ghana MOH, the GHS, and other nutrition stakeholders to strengthen nutrition programs and services and integrate them into the Ghanaian health system (FHI 360 & FANTA, 2014). As of November 2013, 1023 facilities provide CMAM services, 15,025 children treated for severe acute malnutrition, 83 facilities provide NACS services; although the above actions are not HIV-specific. The project facilitated the integration of NACS into routine HIV and TB services, the national nutrition policy, the strategic plan for HIV, TB strategic plans, and the district health management information system; 5533 people living with HIV received supplementary or therapeutic food, 18,688 people living with HIV received nutrition assessment and counseling, and 132 tutors trained on updated curriculums.

The CRS and OICI efforts

The CRS and OICI had implemented nutrition interventions activities for PLHIVs in the last five years. The CRS's Nutrition Support Program was part of the US Title 2 food program, and ended in 2008. It was implemented in five regions (upper east, Ashanti, Volta, eastern, and central) covering 44 districts and over 3056 PLHIVs/beneficiaries. The program provided wheat, wheat-soy blend, and vegetable oil aimed at meeting 20% of daily nutritional needs of the PLHIV and their families. In addition, cost of transportation and packed lunches were given to every PLHIV who attended their monthly meetings. The objectives of the meetings among others were to provide health education, distribute rations, and monitor nutritional status. Assessment and continuous monitoring of nutritional status was not carried out by this project.

The OICI implemented a similar nutrition intervention from 2004 to 2010 covering comparable number of PLHIV in the Greater Accra, Ashanti, eastern and western regions of Ghana. The total budget for the program was five million USD(US $5,000,000) and was funded by the USAID Food for Peace program and the USAID Ghana Mission. The NGO provided vegetable oil, soy fortified sorghum, and soy fortified wheat (whole package was 25 kg worth) to PLHIV and their family, as well as orphans and vulnerable children on a monthly basis. The quantity was reduced to 15 kg near the end of the program. This was done to prevent the flocking of beneficiaries who were exiting the CRS program (CRS provided 15 kg worth of food) to join the OICI.

OICI, the CHAG, CRS, and other stakeholders are currently working with the GAC and DANIDA to implement a home-based care program for PLHIV in the eastern, central, and western regions of Ghana. This program is nutrition sensitive, not nutrition specific, and entails among other things training on comprehensive institutional and home-based care and support as well as strengthening volunteer systems in the communities.

Perspectives of the NAP+ and families affected by HIV

The secretariat for the Ghana network of PLHIV and AIDS, also referred to as NAP+, was established in 2006 to advocate for the rights of PLHIV in Ghana, create an enabling environment for PLHIVs to gain active involvement and participation in the national response, improve the quality of life of all PLHIVs, and prevent further spread of infection. These objectives were envisioned to be achieved through information sharing and capacity strengthening, advocacy for free universal access to treatment for opportunistic infections, antiretroviral drugs and their availability at care and support centers, and strengthening alliances and partnerships with relevant bodies. Almost a decade after the formation of NAP+, stigma and discrimination against PLHIV continue to exist. Some members of the network complained that very limited interventions have been directed at their welfare particularly nutrition.

At its launch, the GAC provided 45,000 Ghana cedis to support NAP+ in their activities and also assisted 32 PLHIVs associations with 3500 Ghana cedis each. Our interactions with the executive members of NAP+ revealed that their members have benefited from two nutrition assistance initiatives since 2005. These were implemented by CRS and OICI. While acknowledging that the support was adequate to cater for the nutritional needs of one beneficiary and six other family members, only 30,000 of the over 200,000 PLHIV were covered by the interventions. About 80% of the over 200,000 PLHIV in Ghana do not work and thereby find it very difficult to feed themselves (president of NAP+). The members of NAP+ that we interacted with ranked food before antiretroviral medications (ARV).

> Taking the drugs (especially the second line regimen) without food induces the feeling of hunger even more, leads to dizziness, hallucinations, and loss of concentration. It is sad to note that some of our members even stopped going to the treatment centers for ART and to attend meetings when the food aid by CRS and OICI ended. We believe food is more important than ARVs. (President of NAP+)

A family who benefited from the OICI intervention concurs with their president.

> We have been counseled on the importance of food in combating the disease. But the food aid we are getting now is insufficient to combat the effects of the drugs we take, not to talk of the disease. The little food distributed to us is not enough to satisfy our hunger. (Family affected by HIV, Eastern Region)

Perspectives from selected HIV/TB treatment centers

The various treatment centers visited acknowledge the critical role nutrition plays in the care and support of PLHIV.

> Nutrition is very key to the efficacy of the treatment. Without proper nutrition, some PLHIV become anemic and even quit medication, as the drugs really demand food. (Healthcare Service Provider, Eastern Region)

The ranking of Nutrition Interventions among other priorities by stakeholders

One opportunity this exercise noted is the fact that all the stakeholders' spoke highly of nutrition interventions. It was ranked first by all but one of the 17 stakeholders interviewed. Food support, some stakeholders expressed, has an added advantage of being used as an incentive for PLHIV to continue to access care and treatment and remain adherent to their drugs.

Even though this recognition was almost unanimous among the stakeholders, there was disconnect between the prioritization and the actual programs/activities being implemented on the ground. Nutrition intervention activities are inadequate. Only 30,000 of the over 200,000 PLHIV (80% of whom do not work and thereby find it very difficult to feed themselves and some 40% of whom, based on a technical report (WFP & GAC, 2011), require social protection interventions) ever received food assistance interventions nationally.

Discussion

Ghana's response to the HIV epidemic has been centered on three thematic areas: prevention, treatment, care; support; and mitigation of socioeconomic effects of HIV. Nutrition is nested under treatment and care. Given that life-saving therapy is still beyond the reach of many PLHIV in Ghana,[1] other complementary approaches for improving the health of HIV-infected persons are needed. Optimal nutrition is one of those strategies. This paper reveals that food and nutrition assistance to PLHIV in Ghana is inadequate. The few documented food and nutrition assistance actions in the country covered an insignificant proportion of those in need.

Issues in Ghana's nutrition and food support response to HIV and AIDS

This analysis of the food and nutrition interventions being implemented by various stakeholders revealed multiple problems. These gaps as well as their possible attributable factors are presented in Table 1.

First, the estimated population of PLHIV (including those with AIDS) in Ghana is approximately 250,000 (NACP, 2011). Based on an average household size of four members (Ghana Statistical Service, 2011), this means that altogether one million people are affected. A nationwide survey involving a nationally representative sample of 1745 households affected by HIV conducted in 2011 provides data on food insecurity among HIV-affected households. The survey found 16% of the 1745 households to be food insecure (WFP & Ghana AIDS Commission, 2011). Relating this prevalence to the entire number of households affected by HIV in Ghana, take the numbers of households who are food insecure in Ghana to about 5979; translated into a total of 23,916 food insecure persons affected by HIV. Given the limited nutrition/food assistance activities/ interventions by the various stakeholders, there is a huge gap regarding meeting the nutrition needs of the food insecure persons affected by HIV. In fact, after the discontinuation of the FFP program in 2010, few PLHIV in Ghana have benefited from any organized food assistance initiative (FHI 360 & FANTA, 2014).

Second, the current state of funding for nutrition intervention activities is inadequate and not sustainable. Also, the few nutrition-specific and nutrition-sensitive policies/ guidelines developed have not yet been fully articulated. It is instructive to note that undernutrition is a nationwide problem and not just peculiar to persons affected by HIV. Despite a number of positive political and strategic steps already taken by Ghana, Ghanaians experiences high rates of maternal and child undernutrition including moderate to severe micronutrient deficiencies. About 10% of Ghanaians are born with a low birth weight and close to 30% of children under five are stunted (GSS, GHS, & Macro-International, 2009). The WFP reported, in 2009, that about 5% of Ghanaians were food insecure and an additional 9% at risk of food insecurity (WFP, 2009).

Although nutrition remains a recognized development challenge, and proven interventions exist, nutrition over decades has retained low priority on the national policy agenda. This is evidenced by the fact that it took Ghana over half a century after her independence to draft her first ever national nutrition policy. However, several government-endorsed documents on nutrition programming and legislation on specific aspects of nutrition do exist (Brantuo et al., 2009; Ghartey, 2011).

A programming gap identified relates to lack of thorough planning. The respondents from the CRS and the OICI noted as weaknesses of their programs, their failure to put in place measures to ensure that food handed out is not abused by recipients/beneficiaries. The programs by the CRS and OICI did not develop exit strategies for the beneficiaries. This put the implementers in a very difficult situation when it came to weaning beneficiaries off. To ensure sustainability, there is a need to establish a system that empowers PLHIV to produce their own food or engage in other income generating activities. The Ghanaian government needs to integrate both nutrition-specific and nutrition-sensitive interventions into their programming. Nutrition-sensitive approaches, such as using poverty reduction strategies, and social protection policies to sustain livelihoods could be integrated into approaches that seek to improve food security and nutrition of HIV-affected persons. Ghana's livelihood empowerment against poverty (LEAP) program could be expanded to cover particularly food insecure PLHIV or female-headed households.

Table 1. A Matrix highlighting problems n Food and Nutrition Support for PLHIVs in Ghana.

Problems to be addressed	Possible attributable factors	Strengths/opportunities	Actions required
Current state of funding for nutrition intervention activities is woefully inadequate and not sustainable.	Even though there is a plethora of scientific evidence on the importance of nutrition in HIV care and support, there is a dearth of advocacy by stakeholders for funding to implement nutrition interventions. There is the absence of national funding; only external funding is being sourced; this is not sustainable, and success rates of such endeavors are also low. Poor representation of the nutrition agenda in important fora that discuss resource mobilization and allocation for HIV and AIDS.	There is almost an overabundance of scientific evidence on the role of nutrition in HIV; the evidence needs to be used for advocacy. There is broad acknowledgement by stakeholders that nutrition is very important in HIV care and support.	Show of urgency in active resource mobilization. One such resource mobilization drive will be to make a case for nutrition in national grant writing/fund seeking activities. To ensure sustainability, there is the need to establish a system to empower PLHIV to produce their own food or engage in other income generating activities. Explore the possibilities of using local food supplements to help improve nutritional status of PLHIV and their families. Explore the possibility and feasibility of scaling up the food by prescription project of US government to cover the whole country.
Limited coverage of nutrition in HIV services including counseling, assessment, and food assistance. No routine/periodic nutrition assessment at the various treatment centers.	Inadequate human resource capacity for nutrition and HIV/AIDS services. Lack of basic equipment for assessment of nutrition status. Few nutritionists have been trained in nutrition and HIV/ADIS.	FANTA II, GHS, NACP plan to develop training curriculum and materials as well as counseling materials for use in the country. Various partners, e.g., UNICEF, GAC are willing to support the initiative.	Capacitate health workers to be able to conduct nutrition assessment, education and counseling of PLHIV. Need for the formation of a Nutrition Technical Working Group that provides guidance on various nutrition issues relating to PLHIV. Such a group should be charged to develop technical guidance notes on how to undertake food and nutritional assessment at the home and clinic levels, among other things.

(*Continued*)

Table 1. (*Continued*).

Problems to be addressed	Possible attributable factors	Strengths/opportunities	Actions required
Lack of thorough planning seems to precede the implementation of nutrition interventions. Neither of the nutrition interventions implemented by the CRS and the OICI put in place measures to ensure that food handed out to beneficiaries are not abused. They also did not develop exit strategies for the beneficiaries, and so found it very difficult to wean off beneficiaries at a point. Currently there are virtually no mechanisms to check and ensure that food given out is not being abused (sold) by beneficiaries. Inadequate coordination and monitoring of nutrition and HIV/AIDS interventions. Little harmonization of nutrition messages being communicated by government and private/NGO agents.	Little experience on how to sustainably manage a food and nutrition program without creating a food dependency syndrome. There is no up to date national database for various stakeholders, partners, and the data being collected on nutrition and HIV in the country. The weak monitoring systems at both facility and community levels also contribute to the gap.	Stakeholders' acknowledgement of their weaknesses and limited experience in sustainable management of food assistance interventions gives enough room for external assistance in capability training.	Food distribution responsibilities need to be well articulated before initiating the food assistance interventions. To ensure equity in distribution, there is the need for partnerships and coordination with other agencies that are providing nutrition support to PLHIV in the country. Improve collaboration among agencies providing nutrition and food support for PLHIV. NAP + should be empowered to play the role of a civil society/pressure group. Institution of effective monitoring systems to ensure that nutrition components of care and support programs are not abused by both beneficiaries and implementers.

The other important gap identified was the fact that routine/periodic nutrition assessment is an optional exercise at the various treatment centers. This exercise uncovered that only weight measurements were sometimes taken. A desk review on the integration of nutrition indicators into Ghana's HIV and tuberculosis monitoring and evaluation reveals that nutrition indicators as of 2013 were not being collected through the system managed by the NACP (Owusu-Amoako, 2013).

Conclusions

Even though stakeholders unanimously viewed nutrition as critically important in the HIV-prevention care cascade, there was disconnect between this theoretical prioritization and the actual actions on the ground. Nutrition intervention activities were found to be inadequate. Some 40% of all PLHIV on ART require some form of social protection interventions, and about 80% of the over 200,000 PLHIV in Ghana find it very difficult to feed themselves, yet only 30,000 PLHIV are ever covered by food assistance interventions nationally. Stakeholders blame the very low coverage on lack of funding. Nutrition indicators are yet to be collected as part of the system setup by NACP for routine data collection.

A more holistic and sustainable approach to addressing this problem will be for the government to integrate both nutrition-specific and nutrition-sensitive interventions into HIV programming. Nutrition-sensitive approaches, such as using poverty reduction strategies, social protection policies to sustain livelihoods could be integrated into approaches that seek to improve food security and nutrition of HIV-affected persons. In particular, Ghana's LEAP program could be expanded to cover particularly food insecure PLHIV households.

Acknowledgments

We are grateful to all the institutions and their staff who participated in this exercise. This paper has benefited from an earlier work commissioned by the United Nations WFP, which sought to document and quantify the gaps in terms of food and nutrition support to PLHIV and affected populations in Ghana.

Note

1. The percentage of eligible adults and children who received antiretroviral therapy in 2009 was 30.5 and 14.8%, respectively (NACP, GHS, & MOH, 2010).

Notes on contributors

Amos Laar has demonstrated research interests spanning HIV, maternal, infant, and young child nutrition. He has experience in planning, implementing and disseminating community-based research. He has been a Co-Investigator of ten successful research grants at the University of Ghana, and has published 17 articles in refereed journals.

Angela El-Adas is the Director-General of the Ghana AIDS Commission, the highest policy making and coordinating body in Ghana on HIV issues. She has led research into sexual and reproductive health needs of vulnerable groups and surveillance among populations most affected by HIV.

Richard N. Amenyah, is the Technical Director of the Ghana AIDS Commission. His research portfolio includes HIV patient monitoring, mapping and size estimation and behavior surveillance among key affected and vulnerable population to HIV infection among others.

Kyeremeh Atuahene is the Director of Research Monitoring and Evaluation at the Ghana AIDS Commission.

Elizabeth Asare, A nutrition scientist with particular interest in HIV, maternal and young child nutrition.

Eric Y. Tenkorang is Assistant Professor of Sociology at Memorial University of Newfoundland, St. John's, Canada. His research has broadly examined the social and cultural determinants of health among at-risk and vulnerable populations in low-income settings

Matilda Laar is a PhD candidate at McGill University. Her background is in international and public health nutrition. Research interests focus on maternal and child undernutrition and effect of maternal HIV on child nutrition in rural and indigenous populations.

Andrew Anthony Adjei, A Professor of Immunolology at the University of Ghana Medical School, whose research covers Malaria, Hepatitis, HIV, and other infectious diseases.

Isabella Quakyi is a health researcher and Public Health Practitioner. Her research contributions span malaria immunology, immunopathology and vaccine development. Her other interests include host parasite relationships on malaria, measles, HIV and AIDS, research capacity building and knowledge translation. Has received substantial research funds and has over 70 publications in peer reviewed journals.

References

Brantuo, M., Okwabi, W., Adu-Afuawuah, S., Agyepong, E., Attafuah, T., Brew, G., … Ashong, J. (2009). Landscape analysis of readiness to accelerate the reduction of maternal and child undernutrition in Ghana. *SCN News, 37*, 31–37.

FHI 360 & FANTA. (2014). *Strengthening nutrition in Ghana: A report on FANTA activities from 2007 to 2013*. Washington, DC.

Ghana Statistical Service. (2008). *Ghana living standards survey: Report of the fifth round (GLSS 5)*. Accra.

Ghana Statistical Service. (2011). *The 2010 population and housing census, Ghana*. Ghana: Ghana Statistical Service.

Ghartey, A. (2011). *Nutrition policy and programs in Ghana: The limitation of a single sector approach. Health, nutrition and population (HNP)*. Discussion Paper, 2011. The International Bank for Reconstruction and Development/The World Bank (81p.).

Gillespie, S., & Kadiyala, S. (2005). *HIV/AIDS and food and nutrition security: From evidence to action*. Food Policy Review 7. Washington, DC: International Food Policy Institute.

GSS, GHS, & Macro-International. (2009). *Ghana demographic and health survey 2008*. Accra, Ghana.

Houtzager, L. (2009). Nutrition in HIV: A review. *Benin Journal of Postgraduate Medicine, 11*, 62–75.

Loevinsohn, M. (2009). *HIV/AIDS, food security and rural livelihoods: Understanding and responding*. Food consumption and Nutrition Division Discussion Paper 157. Washington, DC: International Food Policy Research Institute.

Malterud, K. (1993). Shared understanding of the qualitative research process. Guidelines for the medical researcher. *Family Practice, 10*, 201–206. doi:10.1093/fampra/10.2.201

NACP. (2011). *The 2010 HIV sentinel survey report*. Accra: Ghana Health Service, Ministry of Health.

NACP. (2013). *The 2012 HIV sentinel survey report*. Accra: Ghana Health Service, Ministry of Health.

NACP, GHS, & MOH. (2010). *National guidelines for antiretroviral therapy, Ghana*. Accra, Ghana.

Owusu-Amoako, M. (2013). *Integration of nutrition indicators into Ghana's HIV and tuberculosis monitoring and evaluation system: Summary Report*. Washington, DC.

Panagides, D., Graciano, R., Atekyereza, P., Gerberg, L., & Chopra, M. (2006). *A review of nutrition and food security approaches in HIV and AIDS programmes in Eastern and Southern Africa*. EQUINET Discussion Paper, No. 48. Medical Research Council of South Africa, University of the Western Cape in the Regional Network for Equity in Health in East and Southern Africa.

Rawat, R., Kadiyala, S., & McNamara, P. E. (2010). The impact of food assistance on weight gain and disease progression among HIV-infected individuals accessing AIDS care and treatment services in Uganda. *BMC Public Health, 10*, 316. doi:10.1186/1471-2458-10-316

Shannon, K., Kerr, T., Milloy, M.-J., Anema, A., Zhang, R., Montaner, J. S. G., & Wood, E. (2011). Severe food insecurity is associated with elevated unprotected sex among HIV-seropositive injection drug users independent of HAART use. *AIDS, 25*, 2037–2042. doi:10.1097/QAD.0b013e32834b35c9

The World Bank. (2006). *Directions in development: Repositioning nutrition as central to development, a strategy for large-scale action* (Vol. 13, 272p.). Washington, DC: World Bank Publications.

Villamor, E., Saathoff, E., Mugusi, F., Bosch, R. J., Urassa, W., & Fawzi, W. W. (2006). Wasting and body composition of adults with pulmonary tuberculosis in relation to HIV-1 coinfection, socioeconomic status, and severity of tuberculosis. *European Journal of Clinical Nutrition, 60*, 163–171. doi:10.1038/sj.ejcn.1602281

Weiser, S. D., Fernandes, K. A., Brandson, E. K., Lima, V. D., Anema, A., Bangsberg, D. R., … Hogg, R. S. (2009). The association between food insecurity and mortality among HIV-infected individuals on HAART. *Journal of Acquired Immune Deficiency Syndromes, 52*, 342–349. doi:10.1097/QAI.0b013e3181b627c2

WFP. (2009). *Comprehensive Food Security and Vulnerability Analysis (CFSVA)* (162p.). Accra. Retrieved from http://grad.ochaopt.org/documents/CFSVA_WBGS.pdf

WHO. (2011). *WHO|Progress report 2011: Global HIV/AIDS response* (225pp.). Geneva: World Health Organization. Retrieved from http://www.who.int/hiv/pub/progress_report2011/en/

World Food Programme & Ghana AIDS Commission. (2011, May). *Nationwide study on food security and vulnerability of HIV-affected households in Ghana*.

William, D. (2009). Impact of nutritional supplementation amongst PLHIV in Zambia. *Field Exchange*. Issue 36.

A field assessment of land use systems and soil properties at varied landscape positions in a fragile ecosystem of Mount Elgon, Uganda

Tonny J. Oyana[a,b], Ellen Kayendeke[c], Yazidhi Bamutaze[d] and Danielson Kisanga[e]

[a]*Advanced Geospatial Analysis Laboratory, Department of Geography and Environmental Resources, Southern Illinois University, Mailcode, Carbondale, IL, USA;* [b]*RCHDEE, University of Tennessee at Memphis, Memphis, TN, USA;* [c]*Department of Environment Management, Makerere University, Kampala, Uganda;* [d]*Department of Geography, Geo-informatics and Climatic Sciences, Makerere University, Kampala, Uganda;* [e]*Department of Geography, University of Dar es Salaam, Dar es Salaam, Tanzania*

The purpose of this field study was to identify and characterize land use systems at varied spatial hierarchies, and to differentiate soil properties along varied landscape positions. Key data about existing land use systems and soil properties were systematically collected in 12 sites. Topographic profiles across catchment and geomorphic units were identified to show the relation typical of a mountain ecosystem. We found statistically significant differences ($p < .05$) in soil properties along altitudinal gradients in 52% of the samples and also in land use and land cover distribution characteristics along the slope gradients. The most dominant soils were sandy loam on the hillslopes and clay loams at the footslopes; and most of the cultivated fields were located on slopes ranging from 0 to 17%. The livestock production system was based on a small constructed unit with a capacity to enable feed resources. Soil conservation practices that were evident among the study sites included the use of mulching, hedgerows, intercropping, fallowing, and minimum tillage. Deep empirical knowledge of this system will further advance our understanding on how a fragile ecosystem responds to both natural and human-caused change, and its associated impact on the surrounding environment.

1. Introduction

The complex interaction between land use systems, geomorphology, and soil properties in a heavily settled tropical fragile mountain ecosystems is of substantial value from a livelihood and biogeochemistry perspective. The significance of this study lies in the fact that although there is considerable information and knowledge about different land use systems and soil properties at varied landscape positions, very little is currently known about the variability of such systems, especially in a highly dynamic mountain ecosystem. This study paid particular attention to the types of land use systems and soil properties in this rapidly changing ecosystem that is experiencing significant population growth. The implication of this growth has been observed in the increased demand for agricultural land and settlement expansions over the past 20 years. This has indeed placed enormous pressure on an already fragile ecosystem (Hansen et al., 2013; Kitutu, Muwanga, Poesen, & Deckers, 2009; Mugagga, Kakembo, & Buyinza, 2010, 2011; Myhren, 2007; Namaalwa, van Dam, Funk, Ajie, & Kaggwa, 2013; Okwuosa, 2012; Oonyu, 2009).

It is important to study the variability of land use systems in mountain ecosystem because landscape position in combination with different land use patterns affects the amount of runoff, drainage conditions, soil formation as well as the type of soil erosion (Xue, Cheng, & An, 2013) within the ecosystem. It is vital for farmers to have a good comprehension of these variations and how they influence the productivity of the land in form of soil fertility as well as rainfall conditions (Mugagga et al., 2010, 2011; Okwuosa, 2012); so that they are able to come up with suitable cropping systems and tillage practices at different landscape positions for enhancement of soil and water conservation (Bamutaze, Tenywa, Majaliwa, Vanacker, et al., 2010). Yimer, Ledin, and Abdelkadir (2006) assessed variations in soil properties along three vegetation gradients in the Bale Mountains of south-eastern Ethiopia. Sample plots were randomly located in the selected vegetation zones at different topographic aspects, and soil profiles were taken at each plot as well as information about altitude and landscape positions. In this study, soil properties varied with differences in vegetation communities and topography. Furthermore, Bamutaze, Tenywa, Majaliwa, Vanacker, et al., (2010) studied the infiltration characteristics of volcanic soils on Mt. Elgon Uganda of field sites that were randomly selected across a topographic gradient and they concluded that infiltration rates are correlated to slope gradient.

Mt. Elgon depicts a complex geomorphology hierarchically characterized by a diversity of landscapes, relief forms, and landforms. The outstanding features based on the altitudinal variability are the Caldera, Upper slopes, cliffs and terrace slopes, ridges, and valleys. Although often referred to as 'altitudinal zones', these zones do not strictly follow contour lines. They are a function of the geological processes notably differential volcanicity that was experienced coupled with the subsequent processes of weathering, erosion, deposition, and redeposition as well as landsliding. At a landscape hierarchy, the higher altitudes (2700–3900 masl) on Mt. Elgon hardly have any soils formed on the slopes. Between 1500–2500 masl, there is a significant presence of cliffs and terrace-like slopes that are interspersed with moderately sloping terrain. On level crests, shallow loams have formed over laterite, whereas on the slopes deep red loams occur. The geomorphology of ridges and valleys (at less than 2100 masl) occurs in the southern and south western section of the Mt. Elgon is characterized by angular ridges and valleys with steep slopes (Kitutu et al., 2009; Knapen et al., 2006). Soils are shallow to moderately deep brown clay loams over volcanic rock. The production and farming systems on Mt. Elgon is strongly controlled largely by the spatial variation in climate, geomorphology, geology, and soils.

Land use systems involve the study of socioeconomic and biophysical processes and how they interact to influence the ecosystem function relative to land use, intensity, and management in production system (Figure 1). The land use system of Mt. Elgon is complex; and it is underpinned by the spatial variability a range of biophysical and socioeconomic factors. At the highest landscape hierarchy on Mt. Elgon, a distinct stratification of crop production according to altitude is evident. In the lower to medium altitudinal zones (up to 1800 masl), the landscape typically depicts composite hillslopes in terms of the slopes and land use characterized by scattered planted trees largely eucalyptus, annual and perennial cropping (Bamutaze, Tenywa, Majaliwa, Palesjo, et al., 2010; Bamutaze, Tenywa, Majaliwa, Vanacker, et al., 2010; Okwuosa, 2012). Cultural factors and attitudes also have a strong influence on land use systems, especially among

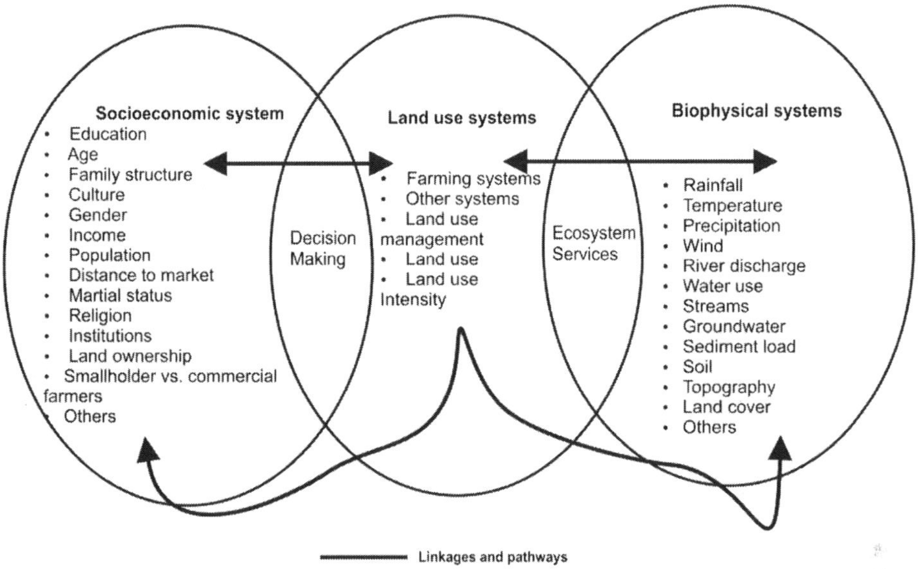

Figure 1. A schematic overview of the land production system in Manafwa catchment, Mt. Elgon ecosystems. There are three systems: the socioeconomic, land use, and biophysical. Each system has a list of variables that play an influential role in determining the dynamics in land production system. The illustrative arrows indicate conceptual linkages and potential feedback loops within the three identified systems.

the Mt. Elgon communities (Barrow, Clarke, Grundy, Kamugisha-Ruhombe, & Tessema, 2002; Myhren, 2007; Oonyu, 2009). Besides, recent studies on land use changes (Hansen et al., 2013; Mugagga et al., 2010, 2011; Okwuosa, 2012) indicated a considerable loss of woodlands and forest, particularly on the steep concave slopes; and a conversion of wetlands to farms (Namaalwa et al., 2013).

Because of the demographic pressure, natural vegetation has been transformed into agricultural landscapes and only remnants of the previous cover can be observed in isolated sites across the hillslopes. At a regional level, Mt. Elgon is classified agro-ecologically as a montane system (Banana–coffee system). There are two major agricultural cropping systems, namely perennial and annual cropping systems. The perennial cropping system is dominated by bananas and coffee hence the 'banana–coffee' classification system. The annuals are dominated by maize, cassava, beans, groundnuts, and potatoes among others. However, it is important to note that a typical annual or perennial farm is difficult to obtain particularly in the medium altitudes; rather they are typically intercrops which can inhabit up to six crops on a single farming field. The last 20–30 years have also seen a gradual but systematic transformation from perennial to annual cropping. The landholding in the low and high altitudinal zones is approximately 1 and .5 ha, respectively. The stream/river valleys on Mt. Elgon show a clear crop gradient with tomatoes, cabbages, onions, and cowpeas grown closer to the river bank. It is estimated that small-scale uniform farm land covers 47% of the total landscape.

Motivated by this background, we collected additional terrain stratified field observations' data about the land production system in Manafwa catchment, Mt. Elgon. The purpose of this field study was (1) to identify and characterize land use systems at varied spatial hierarchies, and (2) to differentiate soil properties along varied landscape positions. We wanted to determine whether there are significant differences in soil properties along altitudinal gradients; and also in land use land cover (LULC) characteristics along slope gradients in Mt. Elgon. Deepening our empirical knowledge about this system will further advance our understanding on how a fragile ecosystem responds to both natural and human-caused change, current pressures placed on, and may help us to better understand other mountain ecosystems in the region.

2. Materials and methods

2.1. Description of the study area

The study area is located in Bubulo County, Manafwa District, Mt. Elgon. This district is situated along the borders in eastern Uganda, to the north the area shares a border with Bududa district, to the east and south is the Republic of Kenya, to the south-west is Tororo district, and Mbale district is in the west (Figure 2). The total surface area is estimated as 451 km^2 with a population of approximately 317,500 people. It is part of the oldest and largest volcano base and has a rich catchment area that serves major rivers, such as Suam, Nzoia, Manafwa, and Sipi. Past vegetation studies done in this area (Hamilton & Perrott, 1981; Howard, 1991; Katende, Ipulet, Rodriques, & Dranzoa, 1989; Langdale-Brown, Osmasten, & Wilson, 1964; van Heist, 1994) have identified four key vegetation zones, which vary by altitude. The zones are lush montane forest (2000–2500 masl), mixed bamboo and low canopy montane forest (2400–3000 masl) with the largest biodiversity, dense shrub, and brilliant wildflowers make up the heath (3000–3500 masl), and moorland and endemic plant species (3500–4231 masl). Some of

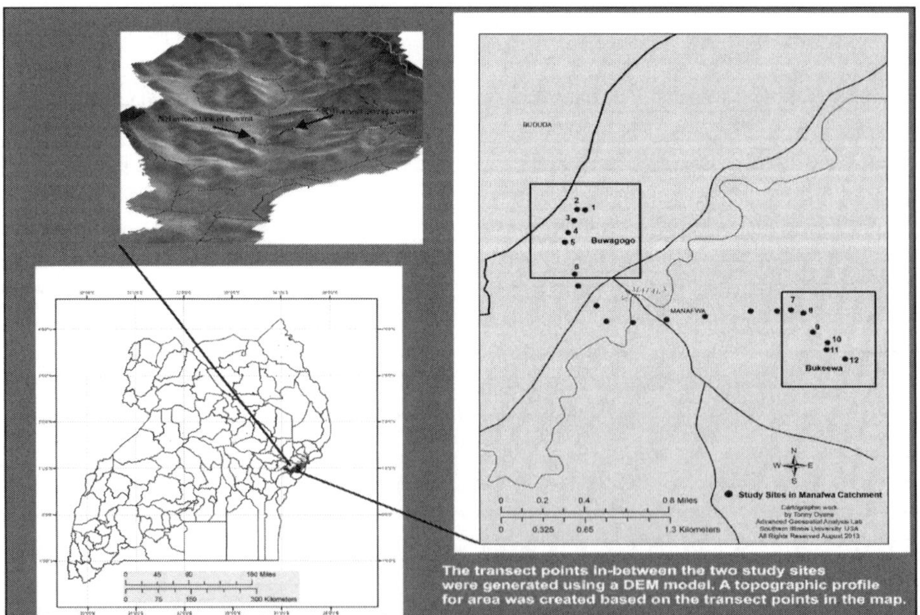

Figure 2. A map showing 12 study sites along the transect points within Manafwa catchment, Mt. Elgon, Uganda. The study sites have been referred to in the text and topographic profile.

common tree species are Elgon Teaks, Cedar trees, Eucalyptus, and Bamboo. The study area is made up of a combination of hills and flat areas with altitude ranging from 1200 to 1700 m. Although this area experiences a bimodal pattern of rainfall, with the wettest months occurring from April to October (van Heist, 1994), most recently, the rains have become less predictable. It now rains throughout the year, but most reports state that the area still receives an average rainfall of 1500 mm per annum.

2.2. *Field study design and data collection*

Land use systems and soil properties were systematically studied in 12 carefully identified sites using Dalrymple's nine slope unit model and Montgomery slope distributions (Dalrymple, Blong, & Conacher, 1968; Food and Agricultural Organization [FAO], 1977; Montgomery, 2001). Line transect surveys were considered the most appropriate spatial sampling method because they enabled the research team to capture principal occurrences of cropping systems and soil properties across altitudinal gradients that were evident throughout this study area. Two line transect surveys were set up at the intersection of lower and upper catchments to systematically study land use systems and soil properties. We focused on two opposite hills situated along the slopes, in Manafwa catchment, Mt. Elgon. The two hills were selected because they represent a good mixture of land use systems, soil–water conservation practices, and soil properties in the catchment. The hills are located in the heart of agricultural activities that lie near River Manafwa. Smallholder farmers from in the area rely heavily on this river as a major water resource for their subsistence needs.

A field study was conducted between 12 June 2013 and 15 June 2013 with fieldwork activities conducted daily from 10 am to 7 pm. The field team consisted of five experts, one graduate student, and two field guides who served as community assistants. Our starting point was on a transect running from a hill that is located in Buwagogo parish, Namuboko Village. Our second study sites were located in Bukeewa parish, Namuboko Village, Manafwa district, which lies at the opposite side of the first hill.

Intensive data collection were done using field instruments, including two measuring tapes, three Juno SD handheld global positioning system (GPS) units with 2–5 m positional accuracy, Nikon D5000 12.3 Megapixel Digital SLR Camera with 18–55 mm and an additional great telephoto lens of 55–300 mm, color flagging, two set squares, and a magnetic angle detector.

Key data were systematically collected about existing land use systems and soil properties in 12 sites. Distinguishing characteristics, including physiographic position, topography of surrounding area, microtopography, aspect, slope, elevation, moisture condition, groundwater level, drainage class, root distribution/effective, soil depth, rock outcrop/surface stones, erosion/deposition, and soil texture information were collected following the United Nations' Food and Agricultural Organization (FAO) guidelines and Dalrymple's 9-unit slope model (Dalrymple et al., 1968; FAO, 1977; Montgomery, 2001). Previous work done in this area was also helpful (Bamutaze, Tenywa, Majaliwa, Vanacker, et al., 2010; Claessens, Knapen, Kitutu, Poesen, & Deckers, 2007; Knapen et al., 2006; Mugagga et al., 2010, 2011; Okwuosa, 2012). A Landsat 8 Path 170/Row 59, 21 May 2013 was acquired from the United States Geological Survey website (http://earthexplorer.usgs.gov/) for use during the field work. A classified image consisting of six LULC classes (cropland, water, built-up area, soil/bareland, forest, and grassland) was generated from this image and used in the compilation of slope distributions for the Manafwa catchment.

The location of study sites was recorded with a GPS and mapped into a geographic information system (GIS) (ESRI's ArcGIS 10.1). Direct observations were made and recorded using a data sheet that was designed to capture physical land characteristics. This was complemented with recordings of audio, photos, and video clips. The attributes associated with the study sites were compiled using Microsoft Excel. The photographs, audio, and video clips were transcribed and used in post-field work data processing, analysis, and report writing. The descriptive English language used for recording observations was standardized using expert knowledge, and where necessary the FAO guidelines and Dalrymple's 9-unit slope model were consulted on any contested classification. Maps and topographic profiles were created to show different land use systems and soil properties. Representative photos showing views of the study sites have been included in this paper (Figure 3).

1 Summit/Interfluve, 2 Creep slope, 3 Fall face to creep convex slope, 4 Concave slope to terrace, 5 Steep convex slope 6 Concave slope, 7Alluvial toe-slope, 8 Channel wall, 9 Channel bed

The 9-unit slope model of Manafwa catchment were identified on June 12th, 2013, Mt. Elgon ecosystem, Uganda

Figure 3. Photographic examples of the 9-unit slope model, six of these units were studied in detail and have been referred to in the text.

Primary and secondary data used in this study were subjected to a range of statistical analyses that included descriptive statistics (mean, standard deviation, and percentages), one-way analysis of variance (ANOVA), and *t*-test.

3. Results and analysis

The observed production system was mainly purposeful and consisted of a strong farming system that was characterized by substantial cropping, livestock, and agro-forestry systems. All the field observations and measurements that were taken based on the 9-unit slope model have systematically been summarized in Tables 1 and 2. Apart from cropping which predominately consists of smallholders and a few large-scale holders/commercial farmlands and was evident throughout the landscape, there were other land uses and land covers, which included livestock rearing, tree plantations, settlements, tertiary infrastructure, roads, cultural and spiritual sites, electric lines, boreholes, sports and recreation, stone quarrying, and sand mining

Photographic examples of the 9-unit slope model are given in Figure 3. Topographic profiles created from field data and digital elevation model (a 15-metre resolution DEM) is presented in Figures 4 and 6. The topographic profile shows illustrative examples of farming systems that were apparent along the 9-unit slope model. In Figure 5, there are multiple summits with unique distinguishing features; however, the land use patterns (farming systems) and soil properties identified from the study sites can be generalized to the larger area (Tables 3–6). Figure 6 shows the topographic profile of a micro-catchment.

The livestock production system had a small constructed unit that enabled feed resources including fresh grass, elephant grass, crop residues, silage, and animal protein. Common livestock includes cattle, poultry, goat, and sheep. This is done primarily for subsistence needs. Major food crops grown were maize, beans, bananas, cassava, pineapples, potatoes, and millet. Many farmers are engaged in growing crops for commercial purposes especially coffee. Many of the observed farms generally had annual, perennial crops, or both. There was significant evidence of intercropping and mixed farming practices with the animals under zero grazing. Some of the farms have a complete agro-forestry system, where the trees (multipurpose trees) are purposely grown to provide green manure or serve as strong wind breaks, land boundaries, pest repellant, and source of nutrients. In other cultivated fields, crops were partially harvested, had no weeds, crop residues or leaves were left behind in the field to act as organic manure.

Field-observed LULC classes consisted of mainly cropland, agro-forestry, fallow/idle cropland, developed/open space, water/river channel, pastureland, natural vegetation (bush, grassland, and wooded grassland), mixed forest (Cypress, Pine or other exotic species, bamboo, Eucalyptus, indigenous trees), tree plantations, herbaceous wetland, and bare land. The most dominant plant species in this area were *Manihot esculanta*, *Impomea batata*, *Cynodon dactylon*, *Bidens pilosa*, *Lantana camara*, *Pennistum purpureum*, *Commelina spp.*, *Dioscorea rotundata*, *Oxalis latifolia*, and *Coffea Arabica*. The tree crown covers range from dense, moderate, to scattered forests.

Figure 7 presents the slope distribution in zones of a 2013 LULC for the Manafwa catchment. The figure indicates that most LULC categories occurred in slope between 0 and 8% with cropland being the largest in size, this was followed by the second slope category of 8–17%. Cultivated fields were also located along slopes that range from 17 to 52%; and only three LULC classes (cropland, forest, and grassland) with slope ranging from 37 to 52% had some agricultural activity or land cover. Surprisingly, farming

Table 1. Transect I is located in Buwagogo Parish, Namuboko Village.

Site #	1	2	3	4	5	6
Physiographic position	Summit	Creep slope	Fall face to creep convex slope	Concave slope to terraces	Steep convex slope	Concave slope
Topography of surrounding sub-county	Mountains	Mountains	Mountains	Mountains	Mountains	Mountains
Microtopography	Flat to gently sloping	Gently sloping	Gently sloping	Terraces	Convex creep slope to fall face	Footslope
Aspect	290°	210°	210°	266°	38°	0°
Slope	Class 2 (gently sloping 2–6%)	Class 3 (sloping 6–13%)	Class 3 (sloping 6–13%)	Class 5 (steep 25–55%)	Class 5 (steep 25–55%)	Class 2 (gently sloping 2–6%)
Elevation	1414 m	1390 m	1373 m	1342 m	1313 m	1249 m
Moisture condition	Dry to moist soil	Dry to moist soil	Dry to moist soil	Dry to moist soil	Dry to moist soil	Dry to moist soil
Groundwater level	Deep	Deep	Deep	Deep	Deep	Deep
Drainage class	Class 5, somewhat excessively drained	Class 6 excessively drained	Class 5, somewhat excessively drained	Class 5, somewhat excessively drained	Class 6, excessively drained	Class 4, well drained
Root distribution/ Effective soil depth	Shallow, sandy loam textured soils	Shallow sandy loam soils	Moderately deep sandy loam soil	Moderately deep clay loam soil	Moderately deep to shallow sandy loam soil	Deep clay loam soil
Rock outcrop/ Surface stones	Class 4, extremely rocky	Class 4, extremely rocky	Class 0, no rocks or few rocks	Class 0, no rocks or few rocks	Class 1, fairly stony	Class 0, no rocks or few rocks
Erosion/Deposition	Excessive water erosion	Excessive water erosion	Minimum water erosion	Minimum water erosion, interrill and rill erosion	Minimum water erosion, interrill and rill erosion	Minimum water erosion, mostly splash
Soil classification	Entisols/Inceptisols	Entisols/ Inceptisols			Entisols/Inceptisols	

Source: Compiled from Field Observations and Standardized Using the United Nations' Food and Agricultural Organization (FAO) guidelines and Dalrymple's 9-unit slope model (Dalrymple et al., 1968; FAO, 1977; Montgomery, 2001).

Table 2. Transect II is located in Bukeewa parish, Namuboko village.

Site #	1	2	3	4	5	6
Physiographic position	Summit	Convex creep slope fall face	Convex creep slope toward fall face	Very steep slope toward fall face	Concave very steep slope	Very steep convex creep to fall face slope
Topography of surrounding sub-county	Mountains	Mountains	Mountains	Mountains	Mountains	Mountains
Microtopography	Flat to gently sloping	Steep slope	Steep slope	Steep slope	Concave steep slope	Very steep slope
Aspect	290°	230°	331°	331°	320°	293°
Slope	Class 2 (gently sloping 2–6%)	Class 2 (gently sloping 2–6%)	Class 5 (steep sloping 25–55%)	Class 5 (steep sloping 25–55%)	Class 5 (steep sloping 25–55%)	Class 5 (steep sloping 25–55%)
Elevation	1622 m	1600 m	1518 m	1342 m	1534 m	1454 m
Moisture condition	Dry to moist soil	Dry to moist soil	Dry to moist soil	Dry to moist soil	Dry to moist soil	
Groundwater level	Deep	Deep	Deep	Deep	Deep	Deep
Drainage class	Class 6, excessively drained	Class 6, excessively drained	Class 5, somewhat excessively drained	Class 6, excessively drained	Class 4, well drained	Class 6, excessively drained
Root distribution/ effective soil depth	Shallow sandy loam textured soils	Shallow sandy textured soils	Shallow sandy loam textured soils	Shallow sandy textured soils	Deep sandy clay loam textured soils	Shallow sandy textured soils
Rock outcrop/ Surface stones	Class 4, extremely rocky	Class 4, extremely rocky	Class 0, no rocks or very few rocks	Class 1, fairly stony	Class 1, fairly stony	Class 4, extremely rocky
Erosion/ Deposition	More of splash, minimum open land surface	Severe splash, interrill and rill erosion	Severe splash, moderate interrill and rill erosion	Severe splash, moderate interrill and rill erosion	Moderate splash, moderate interrill and rill erosion	Severe splash interrill and rill erosion
Soil classification	Entisols/Inceptisols	Entisols/ Inceptisols				Entisols/Inceptisols

Source: Compiled from field observations and standardized Using the United Nations' Food and Agricultural Organization (FAO) guidelines and Dalrymple's 9-unit slope model (Dalrymple et al., 1968; FAO, 1977; Montgomery, 2001).

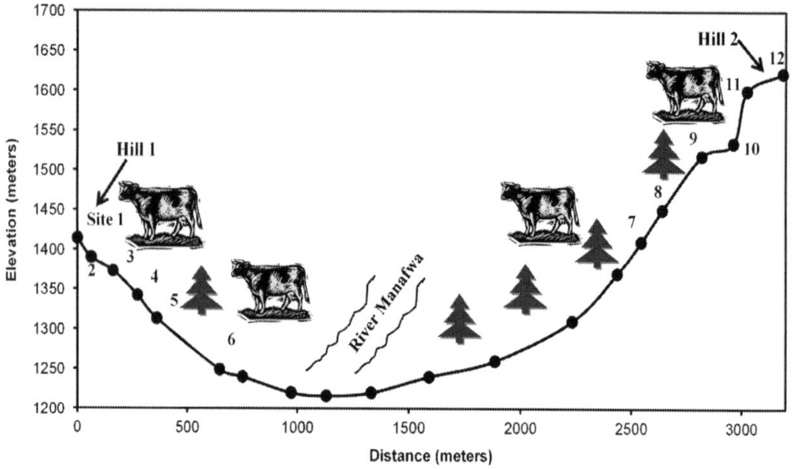

Figure 4. Topographic profile across Manafwa catchment and geomorphic units are identified above the profile and show the relation typical of a mountain ecosystem. The trees and cows represent different human activities (a detailed description is given in Section 3.2) occurring along the cross-section profile, such as zero grazing and agro-forestry activities. The distinguishing characteristics of the 12 study sites have been described in the text.

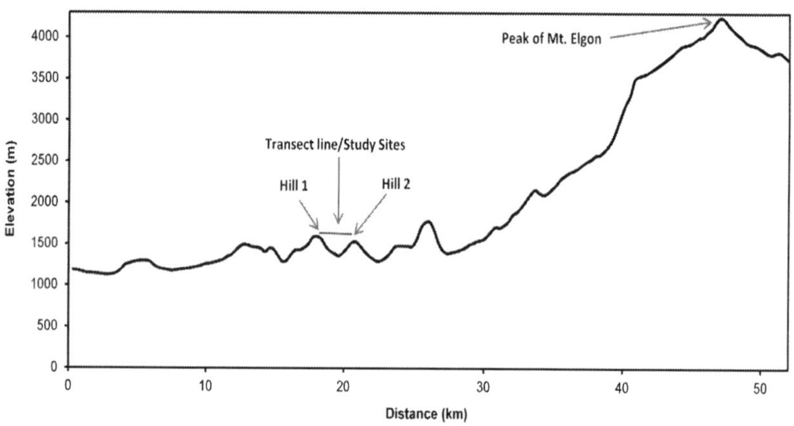

Figure 5. Topographic profile extending from the lower bottom of the catchment to the peak of Mt. Elgon. The location of the transect line/study sites is shown. There are multiple summits with unique distinguishing features; however, the land use patterns and soil properties identified from the study sites can be generalized to the larger area.

occurs even on steep to very slopes (concave and convex steep), which represents approximately 2.5 km^2 of the catchment.

The LULC attributes of Manafwa catchment were divided into six slope positions (Figure 7). Statistical analyses were performed to test the influence of topographic parameters on soil properties and LULC distributions and characteristics using *t*-test and a ANOVA. Some of the key soil properties (synthesized in Table 4) was obtained from five published reports (Bamutaze, Tenywa, Majaliwa, Palesjo, et al., 2010; Bamutaze,

Table 3. Geomorphology and dominant soils on Mt. Elgon.

%	Area (Km^2)	Geomorphic unit	Parent material	Classification FAO/ UNESCO	Classification USDA	Productivity	Crops
29	1213	Volcanic rock	Mt. Elgon volcanics	Umbric Andosols	Paleudults and Paleustults	Low-medium	Forest, occasionally food crops
11	456	Volcanic mountain	Mt. Elgon volcanics	Haplic Andosols	Rhodic Eutrandept	High-medium	Arabica coffee
9	358	Volcanic rock	Mt. Elgon volcanics	Umbric Andosols	Paleudults and Paleustults	High-medium	Arabica coffee
5	218	Volcanic mountain	Mt. Elgon volcanics	Folic Histosols	Histosols	Nil	Nil
11	477	Buganda surface	B. C granite	Rhodic Ferralsols	Kandiudalfic Eutrudox	Medium	Cotton, Arabica coffee
3	135	Buganda surface	B. C granite and loam amphibolites	Rhodic Nitisols	Kandiudalfic Eutrudox	Medium	Cotton, Arabica coffee
2	86	Buganda surface	B.·C mica, schists and amphibolites	Rhodic Ferralsols	Kandiudalfic Eutrudox	Medium	Arabica coffee
30	1263	Karamoja plateau	Ancient alluvium and colluvium	Vertisols	Vertisols	Medium	Dry-season grazing, some food crop

Source: Compiled from Langdale-Brown et al. (1964); FAO (1977); Bamutaze, Tenywa, Majaliwa, Vanacker, et al. (2010); Bamutaze, Tenywa, Majaliwa, Palesjo, et al. (2010); and other sources.

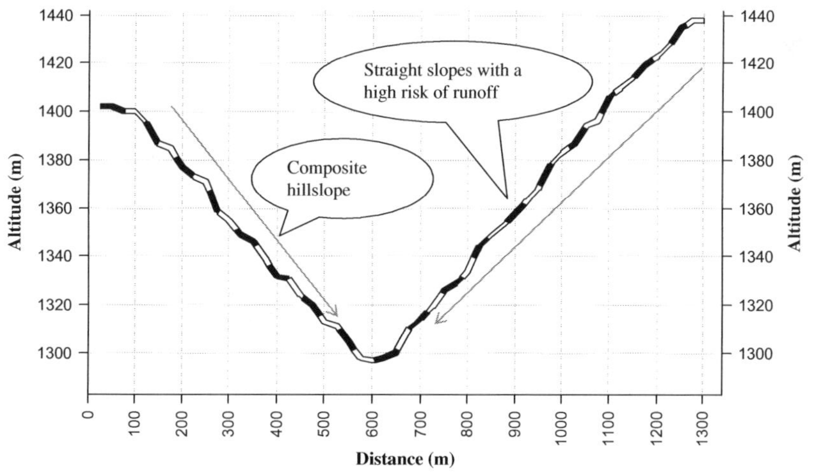

Figure 6. A typical cross-section of a micro-catchment on Mt. Elgon derived from a Digital Elevation. It shows a high diversity of micro-relief and risk to a range of surface processes and all the hillslopes in this altitudinal zonation are cultivated.

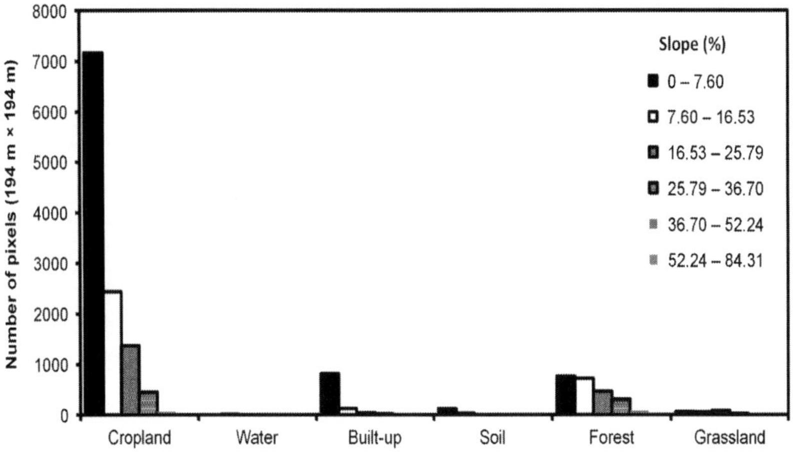

Figure 7. Slope gradient distributions in zones of a 2013 land use and land cover attributes for Manafwa catchment.

Tenywa, Majaliwa, Vanacker, et al., 2010; Mugagga et al., 2010, 2011; Okwuosa, 2012). Mean comparisons were made using Bonferroni–Holm method with $p < .05$. We tested the following two hypotheses.

H_0: There were no significant differences in key soil properties along altitudinal gradients in Mt. Elgon. To test this hypothesis, three mean comparisons were done (Table 5).

Comparison I: 1500 masl vs. 1800 masl

Comparison I indicates that 57% of the key soil properties from averaged sample plots along altitudinal gradients in Mt. Elgon were statistically significant.

Comparison II: 1100 masl vs. 1500 masl

Comparison II indicates that 71% of the key soil properties from averaged sample plots along altitudinal gradients in Mt. Elgon were statistically significant.

Comparison III: 1100 masl vs. 1800 masl

Comparison III indicates that only 30% of the key soil properties from averaged sample plots along altitudinal gradients in Mt. Elgon were statistically significant.

Based upon the statistical results presented in Table 5, it is quite evident that only 52% of the key soil properties in three comparisons were statistically significant. The highest variations were apparent especially in Comparison II (1100 masl vs. 1500 masl) and lowest in Comparison III (1100 masl vs. 1800 masl). However, there were no significant differences in exchangeable potassium in soils in the three comparisons. Revisiting the second hypothesis, it is quite clear that we have mixed statistical results; in most of the

Table 4. Mean rates of selected soil properties at different altitudinal zones on Mt. Elgon.

Altitude (masl)	SOM (%)	N (%)	P (ppm)	K (Cmo·kg^{-1})	Sand (%)	Silt (%)	Clay (%)	Textural class
1100	2.86 ± .87	.12 ± .04	19.9 ± 16.9	.75 ± .02	40.3 ± 13.64	24.6 ± 6.16	35.02 ± 8.63	Clay loam
1500	2.6 ± .48	.16 ± .02	7.0 ± 5.7	.69 ± .41	27.6 ± 3.92	20.54 ± 3.7	51.5 ± 5.2	Clay
1800	4.14 ± .92	.4 ± .15	11.4 ± 5.44	1.1 ± .77	44.5 ± 6.88	17.2 ± 5.29	38.3 ± 7.52	Clay loam

Source: Compiled from Mugagga et al. (2010); Mugagga et al. (2011); Bamutaze, Tenywa, Majaliwa, Vanacker, et al. (2010); Bamutaze, Tenywa, Majaliwa, Palesjo, et al. (2010); Okwuosa (2012).

Table 5. *t*-Test results for altitudinal zone comparisons.

	SOM	N	P	K	Sand	Silt	Clay
1500 masl vs. 1800 masl							
t-Statistics	5.14	5.49	1.93	1.63	7.39	1.76	5.0013
p-Value	.0001*	.0001*	.066	.118	.0001*	.093	.0001*
1100 masl vs. 1500 masl							
t-Statistics	.91	3.098	2.506	.506	3.099	3.153	5.666
p-Value	.375	.005*	.0201*	.618	.0052*	.0046*	.0001*
1100 masl vs. 1800 masl							
t-Statistics	3.502	6.248	1.659	1.574	.952	1.365	.332
p-Value	.002*	.0001*	.1114	.1297	.3513	.186	.993

Note: Degrees of freedom 22.
*Statistically significant at $p < .05$.

comparisons, we have rejected the null hypothesis, while in others, we have confirmed the null hypothesis.

> H_1: There were significant differences in LULC attributes along the slope gradients in Mt. Elgon.

To test this hypothesis, LULC attributes were summarized by six slope positions using the zonal statistics in ArcGIS. A one-way ANOVA was conducted to test the mean differences of LULC characteristics in six slope positions of the catchment under the null hypothesis of their being equal. F_{stat} was 2.60 and $F_{critical}$ with degrees of freedom (5, 30) was 2.534 at $p < .046$. Given that F_{stat} was greater than $F_{critical}$, we rejected the null hypothesis. This implies that there are significant differences in LULC attributes in the six slope positions.

Table 6. Landscape and farming system characterization.

Landscape characteristic	Highland	Lowland
Altitude	1200–3000 masl	Less than 1200
Population density (persons/km^2)	501–600	201–300
Climate	Humid	Subhumid
Crops	Perennial (banana, coffee)	Few perennial (banana, coffee)
	Annual (maize, beans, cassava, Irish potatoes, vegetables)	Mostly annual (maize, beans, cotton)
Crop-livestock interaction	Intensive system	Extensive system
Soil productivity constraint	Soil erosion, nutrient mining	Nutrient mining
Soil fertility level	Low to medium	Medium to high
Soil and water conservation	Integration of structures and vegetative methods	Vegetative methods
Soil fertility improvement	Use inorganic and organic fertilizers	Use mostly organic fertilizers

Source: Mugagga et al. (2010); Mugagga et al. (2011); Bamutaze, Tenywa, Majaliwa, Vanacker, et al. (2010); Bamutaze, Tenywa, Majaliwa, Palesjo, et al. (2010); Okwuosa (2012) and others.

Mean comparisons showed statistically significant differences ($p < .05$) in water vs. forest, water vs. grassland, soil vs. forest, and forest grassland attributes. The mean comparisons among the remaining LULC attributes were statistically insignificant.

A meta-analysis of selected soil properties from a compendium of five experimental studies (Bamutaze, Tenywa, Majaliwa, Palesjo, et al., 2010; Bamutaze, Tenywa, Majaliwa, Vanacker, et al., 2010; Mugagga et al., 2010, 2011; Okwuosa, 2012) in the annual and perennial cropping systems on Mt. Elgon (Tables 3–6) as proxies of the land quality at three altitudinal zones does not give a definitive landscape-based trend. Of critical importance is the fact that the observed amounts of most soil properties are above the critical levels for crop growth, even though nitrogen levels have been found to be higher in forest land uses than cultivated fields (Okwuosa, 2012). This dovetails with some farmers who perceive their soils to be fertile despite reports indicating that the soil quality is retrogressing due to soil erosion.

At a finer landscape scale, although in some comparisons of soil properties at the lower, middle, and upper slope positions, we observed statistically significant results in 52% of the samples, we did not find definitive differences, largely due to the intercropping practice but also as a result of the numerous slope breaks. The main threats to cultivation without adequate conservation measures on the Mt. Elgon slopes particularly on marginal and fragile sites are runoff, runon, sediment, and nutrient transportation and slope failures. Soil conservation practices that were evident among the study sites included the use of mulching, hedgerows, mixed cropping of cereals with legumes, intercropping, fallowing, and minimum tillage. Terracing as a form of soil conservation measure was evident in some areas; some farmers used boulders. The slopes ranged from generally flat, gently sloping, steep to very steep. The soil depth in some of the cultivated fields, especially in the summit zone, was approximately .17 m. Overcultivation was also evident illustrating land use intensity across the landscape. Moreover, some farmers rarely applied any manure; they cultivated annual crops two times a year following the pattern of bimodal rainfall season. Human-induced LULC changes appeared to be the most influential in the observed areas.

3.1. Agro-ecological zones of Mt. Elgon

There are about 10 agro-ecological zones in Uganda. The agro-ecological zones are shown in Figure 8, and manifest the biophysical and social economic diversity across the country particularly in terms of soils, slope gradient, altitude, rainfall, dominant crops, and livestock. The most prominent parameter in the delineation of the agro-ecological zones in Uganda is the rainfall amount and rainfall effectiveness in supporting crop productivity. On Mt. Elgon, there are two agro-ecological zones namely, the montane system and the medium altitude intensive banana-coffee system. Both the montane system and the medium altitude intensive banana-coffee system are uniquely located in highland environments in Uganda. The montane system on Mt. Elgon is generally confined to areas beyond 1500 m above sea level. On Mt. Elgon, these areas are dominated by small land holdings growing intercrops of coffee, banana, and a range of annual crops. The rainfall amounts in the montane zones generally exceed 1500 mm annually. Rainfall is distributed in two seasons, but the combination of biophysical conditions limit the growth of some crops like maize to just one season due to the extended growing period. The medium altitude intensive banana-coffee system zone occurs just below the montane system generally in the range of 1000–1600 m above sea level. It is still dominated by bananas and coffee, but there is also a higher intensity of annual crops which

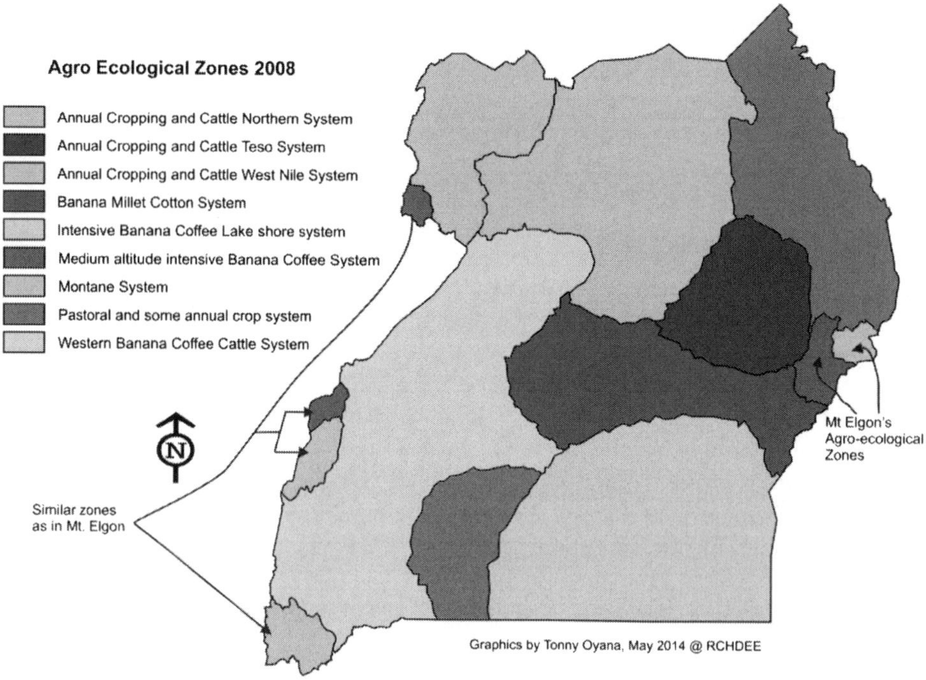

Figure 8. The main agro-ecological zones and Mt's Elgon's two systems are visually illustrated with other similar ones in Uganda. The agro-ecological zones reflect the biophysical and social economic diversity across the country.

are intercropped with the perennials. The two agro-ecological systems in Mt. Elgon are typically positing on soils that have evolved from volcanism. The landscape bearing the two systems is sensitive due to the steep gradients with over 40% being beyond the critical threshold of 30%. However, the soils in many places are also deeper and previous studies (Bamutaze, Tenywa, Majaliwa, Vanacker, et al., 2010) have shown that the soil–water infiltration rates are high. Therefore, the high soil infiltration rates coupled with the high intensity of perennial cropping mitigates runoff and soil loss along the hillslopes.

3.2. Physiographic zones, landscape properties, and land use of Manafwa catchment, Mt. Elgon

(1) The summit/interfluve zone: This is the topmost zone along the hillslope with a conical shape at the zenith. The summit zones are intensively utilized and are more pronounced with agricultural practices. The utilization of the summits for agricultural practices is not surprising since this is one of the most fragmented regions in Uganda owing to the high-population density. The cultivation in this zone is has implications on hillslope processes particularly sediment entrainment and downslope transfer mechanisms, since the slope length component is increased. The summit zones are cultivated with banana plantations, coffee, maize, yams, and cassava. The observed crops were relatively in

poor health. There are soil and water conservation measures manifested by widely spaced grass bands. The common practice in the region is that these grass bands are primarily used as boundaries separating farmer ownership. They nevertheless mitigate runoff on the slope and the grass is also used to feed animals under zero grazing systems in this area. There is limited use of *in situ* SWC, but farmers use NPK fertilizers (nitrogen (N), phosphorus (P), and potassium (K) and organic manure to enhance fertility. Much of the land in this zone has been subjected to soil erosion and there is little horizon differentiation. The soils are sandy and very porous. Rock outcrops are about 3.5 m apart and cover between 50 and 90% of the land surface. Surface sealing and cracking is evident and natural vegetation cover is virtually extinct.

(2) The creep slope zone: This is a zone below the summit, which is transient for surface processes especially runoff. The creep zone is dominated by maize (95%) and cassava cultivation with patterns of mono-cropping and intercropping. The zone also depicts relics of natural forest trees and scattered agro-forestry. The slopes are very steep and water flows rapidly as runoff from the sites. Rock outcrops are about 3.5 m apart and cover between 50 and 90% of the land area. The zone is also exposed to surface sealing and there is evidence of salinity near the surface of degraded soil. There are some terraces serving as SWC measures as well as lower boundary between different farmers and delineation of land tenure ownership.

(3) The fall face to creep convex slope zone: Cultivated fields are dominated by intercrops of banana and coffee, maize, yams, ground nuts cassava, agro-forest trees (mango and avocado), exotic trees, and Eucalyptus trees. There are no bedrock exposures to interfere with tillage. This implies relatively deeper soils. The bedrock exposure is approximately 2%. There were intentional and deliberate SWC practices constituting strip elephant grass.

(4) The concave slope to terraces zone: Cultivated fields are dominated by maize, yams, ground nuts, cassava, agro-forest trees (mango and avocado), few natural tree cover of about 100 years, old tree stumps, copping practices, and there is evidence of well-managed and unique intentional soil and water conservation practices. There is moderately deep clay loam soil caused by deposition from upslope land units trapped by listed crops, trees, and well-maintained terraces and strip cropping system. There is minimum water erosion, interrill, and rill erosion may be due to steep sloped landscape, but the landscape is almost fully covered by crops, agro-forestry, and few natural trees, thus less-intensive erosion. There were some deliberate SWC practices constituting grass bands dominated by elephant grass, accompanied with terraces.

(5) The steep convex slope zone: There is minimum water erosion, interrill and rill erosion may be due to steep sloped landscape, but the landscape is almost fully covered by crops, agro-forestry, and few natural trees, thus less intensive erosion.

(6) The concave slope zone: Cultivated fields are dominated by banana, cassava, coffee, maize and ground nuts; intensive agro-forestry, and scattered *Cassia spp*. Water appears to be removed from the soil readily but not rapidly. The zone appears to retain optimum amount of moisture for plant growth. There are no significant bedrock exposures with only about 2% observed. There is minimum water erosion, mostly splash and the zone is predominantly a deposition area due to the gently sloped landscape and the good vegetative cover.

Terraces appear at the bottom parts of each small farmland plots marking the borders between individual farmers.

(7) The very steep convex creep to fall face slope zone: This zone has evidence of perennial and annual crops on a very steep terrain with poor performance – banana, yams, maize, vegetables, agro-forestry, Eucalyptus, and *Acacia robusta*. Water is removed from the soil very rapidly and there is evidence of soils that are excessively drained, especially on steep slopes and very porous soils. There was sufficient rock outcrop of about 3.5 m apart ranging from 50 to 90% of the land surface area. Zero grazing at the summit was evident.

(8) The concave very steep slope zone: This zone is thickly covered by agricultural crops plus agro-forestry. The crops include maize, banana, eggplant, coffee, mangoes, and guava. There are also some relics of natural forest trees. Water is removed from the soil readily but not rapidly and appears to retain optimum amount of moisture for plant growth after rains or additions of irrigation water. The surface stone coverage was extremely low (about 1%); thus, there is limited interfere with tillage. The stones on the site were 15–30 cm in diameter and 30 m apart. There was no evidence of seriously soil erosion observed due to the good vegetation cover and slope concavity.

(9) The very steep slope toward fall face zone: Cultivated fields are dominated by maize, banana, cassava, coffee, yams, beans most of which have poor health; agro-forestry (guava, avocado, mangoes) with minimum SWC measures. Runoff and soil losses are predominant in the zone due to the steepness of the slopes and soils are very porous. The surface stone coverage is about 11%. The stones are 15–30 cm in diameter separated by about 10–30 m. However, there was no evidence of SWC practices in place. Zero grazing at the site was evident, particularly at homesteads.

(10) The convex creep slope toward fall face zone: Cultivated fields in this zone are dominated by banana, coffee, beans, and eggplants with proper management, yams, and pumpkins. Animal rearing constitutes of zero grazed Friesian dairy cows at the homesteads. The soils are eroded and have little horizon differentiation and are sandy and very porous. There are sandy loam textured soils about 30 cm thick. However, there are intended SWC practices in place. Diversion ditches could be seen which trapped runoff and cracked clay trapped in the ditches.

(11) The convex creep slope fall face zone: It is pure agriculture zone which is cultivated largely with banana, coffee, and yams. Agro-forestry, limited Eucalyptus, and *Grevillea robusta* are also evident. The zone experiences soil erosion. The soils are excessively drained and very porous. Rock outcrops exist and are about 3.5 m covering 50–90% of the land surface area.

(12) The summit/interfluve zone: This zone is characterized by nonefficient cultivation of maize, cassava, sweet potatoes, and beans. There is some evidence of vegetation cover and different type of land use including Eucalyptus, grassland, telecommunication towers, and minimum agro-forestry. Water appears to be removed from the soil very rapidly and there is evidence of excessively drained soils on the steep slopes and very porous soils. There is sufficient rock outcrop (or very shallow over rock) to make all use of machinery impractical. Rock outcrops are about 3.5 m apart or less and cover some 50–90% of the

area. There is human influence, soil fauna, surface sealing/cracking, and salinity. However, there are minimum SWC practices, land surface almost fully covered with either grass or annual crops. Zero grazing at the summit was evident.

4. Conclusions

Topographic profiles across catchment and geomorphic units were identified and show the relation typical of a mountain ecosystem. The distinguishing characteristics of the 12 study sites were identified and reported. We found statistically significant differences ($p < .05$) in soil properties along altitudinal gradients in 52% of the samples and also in LULC attributes along the slope gradients in the study area. These statistical results show differences in soils and LULC characteristics at different elevations depicting the dynamics between physical and human activities across the entire landscape. This was further confirmed in the 12 study sites where we observed the most dominant soils to be sandy loam on the hillslopes and clay loams at the footslopes; most of the settlements are random, but there is more concentration in low-lying areas closer to the road network; and most of the cultivated fields were located on slopes ranging from 0 to 17%.

Furthermore, the most dominant farming practice was intercropping, and some homesteads practiced zero grazing. The livestock production system was based on a small constructed unit with a capacity to enable feed resources including fresh grass, elephant grass, crop residues, silage, and animal protein. The major food crops grown in the catchment were coffee, maize, beans, bananas, cassava, pineapples, potatoes, and millet. Soil conservation practices that were evident among the study sites included the use of mulching, hedgerows, mixed cropping of cereals with legumes, intercropping, fallowing, and minimum tillage. Generally, a few farmers practiced agro-forestry and other soil conservation practices, but strip cropping was minimally practiced. Empirical knowledge obtained from this study can help the farmers and agricultural extensive workers to design and develop appropriate cropping systems and tillage practices at different landscape positions.

Acknowledgment

We wish to thank START International for their funding support under the Grants for Global Environmental Change Research in Africa #Subaward 2013-06.

Notes on contributors

Tonny J. Oyana is a professor at University of Tennessee, Memphis and director of Spatial Analytics and Informatics Core in the RCHDEE Center in the United States. He earned his Doctorate of Philosophy in Geography with a main concentration in GIS at the State University of New York at Buffalo. He has over 20 years of demonstrated research and academic experience grounded in North America, South America, Asia, Europe, and Africa. Dr. Oyana's research interests include Geographic Information Science and Spatial Analytical Methods. Within these areas, he has several substantive foci, including GIS, spatial analysis, visualization and cartographic design, and spatial epidemiology. He works are widely published with over 80 scientific publications including 35+ journal articles, 20+ refereed conference proceedings, and 9 book chapters.

Ellen Kayendeke is an assistant lecturer in Department of Environmental Management at Makerere University, Uganda. She has master's degree in Geo-information science and Earth observation specializing in natural resource management from the Faculty of Geo-information Science and Earth observation (ITC), University of Twente, The Netherlands. Her research interests are in natural resource monitoring through time series analysis and spatial modeling of ecosystem properties.

Dr. Yazidhi Bamutaze is a senior lecturer at Makerere University and coordinator of graduate programs in the Department of Geography, Geo-Informatics, and Climatic Sciences. He holds a PhD in Geography from Makerere University and a Master's degree in Geo-information Science and Earth observation from the Faculty of Geo-information Science and Earth observation (ITC), University of Twente, The Netherlands. His research spans a wide range of issues including natural resource degradation analysis, surface process analysis covering erosion and sediment loading, land use change, and climate change adaptation. He has published over five journal papers and one book chapter.

Danielson R. Kisanga is a lecturer at University of Dar-es-Salaam and coordinator for Master of GIS program in the Department of Geography. He earned his Doctorate of Philosophy in Geography with a main concentration in Geomorphology, GIS, and Remote Sensing at Clark University. His research interests involves applying GIS and Remote Sensing techniques to solve real-world problems associated with soil resources and land use/cover changes in different climatic regimes. He has studied wetland changes in East Africa and land use/cover changes on tropical mountain ecosystems where he was successful at publishing a few journal articles, book chapters, and encyclopedia entries.

References

Bamutaze, Y., Tenywa, M. M., Majaliwa, M. G., Palesjo, P., & Bagoora, F. (2010). Terrain and land use influence on patterns of soil properties in a micro-catchment on Mt. Elgon, Uganda. *African Geographical Review, 29*, 67–90.

Bamutaze, Y., Tenywa, M. M., Majaliwa, M. G., Vanacker, V., Bagoora, F., Magunda, M., ... Wasige, J. E. (2010, February 15). Infiltration characteristics of volcanic sloping soils on Mt. Elgon, Eastern Uganda. *Catena, 80*, 122–130, ISSN 0341-8162. Retrieved from http://dx.doi.org/10.1016/j.catena.2009.09.006

Barrow, E., Clarke, J., Grundy, I., Kamugisha-Ruhombe, J., & Tessema, Y. 2002. *Analysis of stakeholder power and responsibilities in community involvement in forest management in eastern and southern Africa* (Forest and Social Perspectives in Conservation Number 9). Nairobi: IUCN The World Conservation Union Eastern African Programme.

Claessens, L., Knapen, A., Kitutu, M. G., Poesen, J., & Deckers, J. A. (2007). Modelling landslide hazard, soil redistribution and sediment yield of landslides on the Ugandan footslopes of Mount Elgon. *Geomorphology, 90*, 23–35.

Dalrymple, J. B., Blong, R. J., & Conacher, A. J. (1968). A hypothetical nine-unit land surface model. *Zeitschrift für Geomorphologie, 12*, 60–76.

Food and Agricultural Organisation. (1977). *Guidelines for soil profile descriptions* (2nd ed., 66 p). Rome: Author.

Hamilton, A. C., & Perrott, R. A. (1981). A study of altitudinal zonation in the montane forest belt of Mt. Elgon, Kenya/Uganda. *Vegetatio/Plant Ecology, 45*, 107–125.

Hansen, M. C., Potapov, P. V., Moore, R., Hancher, M., Turubanova, S. A., Tyukavina, A., ... Townshend, J. R. G. (2013). High-resolution global maps of 21st-Century forest cover change. *Science, 342*, 850–853. Retrieved from http://dx.doi.org/10.1126/science.1244693

Howard, P. C. (1991). *Nature conservation in Uganda's tropical reserves*. Kampala: Forest Department/Ministry of Environment Protection Uganda.

Katende, A. T., Ipulet, P., Rodriques, R., & Dranzoa, C. (1989). *Birds and woody perennials inventory: Mount Elgon forest reserve. Sustainable development and forest conservation in Uganda* (IUCN Technical Report No. 1). Nairobi: IUCN.

Kitutu, M. G., Muwanga, A., Poesen, J., & Deckers, J. A. (2009). Influence of soil properties on landslide occurrence in Bududa district, Eastern Uganda. *African Journal of Agricultural Research, 47*, 611–620.

Knapen, A., Kitutu, M. G., Poesen, J., Breugelmans, W., Deckers, J., & Muwanga, A. (2006). Landslides in a densely populated county at the footslopes of Mount Elgon (Uganda): Characteristics and causal factors. *Geomorphology, 73*, 149–165.

Langdale-Brown, I., Osmasten, H. A., & Wilson, J. G. (1964). *The vegetation of Uganda and its bearing on land use*. Entebbe: Government Printer.

Montgomery, D. R. (2001). Slope distributions, threshold hillslopes, and steady-state topography. *American Journal of Science, 301*, 432–454.

Mugagga, F., Kakembo, V., & Buyinza, M. (2010). A characterization of the physical properties of soil and the implications for landslide occurrence on the slopes of the Mount Elgon, Eastern Uganda. *Natural Hazards, 52*, 1113–1131. Retrieved from http://dx.doi.org/10.1007/s11069-011-9896-3

Mugagga, F., Kakembo, V., & Buyinza, M. (2011). Land use changes on the slopes of Mount Elgon and the implications for the occurrence of landslides. *Catena, 90*, 39–46.

Myhren, S. R. (2007). *Rural livelihood and forest management by Mount Elgon, Kenya* (Unpublished Msc thesis in development studies). Norwegian University of Life Science, Oslo.

Okwuosa, E. A. 2012. *Effect of land use system on soil carbon and selected soil properties in Mt. Elgon ecosystem, Kenya* (Unpublished master thesis). Agricultural Information and Communication Management, University of Nairobi, Kenya.

Oonyu, J. C. (2009). Conservation education and the attitudes of local communities living adjacent to Mt. Elgon National Park, Uganda. *Applied Environmental Education & Communication, 8*, 153–164.

van Heist, M. 1994. *Accompanying report with the land unit map of Mount Elgon National Park* (IUCN Technical Report, unpublished). Nairobi: IUCN.

van Namaalwa, S., Dam, A. A., Funk, A., Ajie, G. S., & Kaggwa, R. C. (2013). A characterization of the drivers, pressures, ecosystem functions and services of Namatala wetland, Uganda. *Environmental Science & Policy*. Retrieved from http://dx.doi.org/10.1016/j.envsci.2013.01.002

Xue, Z., Cheng, M., & An, S. (2013). Soil nitrogen distributions for different land uses and landscape positions in a small watershed on Loess Plateau, China. *Ecological Engineering, 60*, 204–213.

Yimer, F., Ledin, S., & Abdelkadir, A. (2006). Soil property variations in relation to topographic aspect and vegetation community in the south-eastern highlands of Ethiopia. *Forest Ecology and Management, 232*, 90–99.

Smallholder Perspectives on Soil Fertility Management and Markets in the African Green Revolution

IMOGEN RUBY VANESSA BELLWOOD-HOWARD
Department of Geography, Kings College London, London, UK

The African Green Revolution (AGR) describes a drive by governments and international philanthropic and research organizations to raise agricultural production through Integrated Soil Fertility Management (ISFM) and improved access to input and output markets. This article draws on data from three studies that took place in Northern Ghana as the AGR unfolded between 2008 and 2012, focusing specifically on soil fertility management (SFM) and access to credit markets. It finds that the AGR idea does incorporate some elements farmers find useful, indeed, that they are implementing autonomously, such as use of organic soil amendments. However, the underpinning emphasis on commercialization does not allow room for them to practice site-specific SFM and subsistence mechanisms that they have developed as responses to risk. Some aspects of the AGR are themselves disadvantageous, for example repayment of credit at harvest time, the emphasis on inorganic fertilizer and land titling schemes. Despite advocating farmer choice, the current hegemonic implementation of the AGR makes it unlikely to support farmers' sustainable agroecological practice.

INTRODUCTION

The 21st century has seen a concerted effort by African Governments, international philanthropic and research organizations and private enterprises to bring smallholders into modernized, commercialized agriculture. This African Green Revolution (AGR) focuses on raising production through integrated soil fertility management (ISFM) as well as providing better access to markets upon which to sell such produce (Toenniessen et al. 2008; Sanginga and Woomer 2009; Vanlauwe, Bationo, et al. 2010). The approach is, however, contested, by commentators focused on the ecological and political implications (Altieri 2009; Odame and Muange 2011; Koopman 2012). The start of the current phase of the endeavor could reasonably be seen as the 2006 Abuja fertilizer summit, where African Union member countries pledged to increase fertilizer application rates from an average of 8 kg/ha to 50 kg/ha by 2015 (African Union 2006). The Alliance for a Green Revolution for Africa (AGRA) was founded at this event, marking a new injection of energy into the campaign, and enough evidence has accumulated since to facilitate analysis of the movement. Accordingly, this article draws on data from five years of research in Northern Ghana on soil fertility management (SFM) and maize commercialization. It starts by describing the AGR and the divided opinions on it. Data from three interlinked projects are presented. The implications of each are noted, and the discussion contextualizes the results relating to SFM and markets with reflections on land ownership and economic differentiation. This leads to the conclusion that although the AGR endorses some valuable mechanisms and practices that farmers are already using, such as organic SFM, its overall commercial orientation prevents it from fully meeting all farmers' needs. It also contains elements that have the potential to further disadvantage farmers who prefer not to adopt its hegemonic ideology.

THE AFRICAN GREEN REVOLUTION

The AGR builds on the technologist model of the 1960's Asian Green Revolution (GR). Its main thrust is that institutional and financial reforms are just as important as technological fixes, and that if this is recognized, many of the less advantageous effects that resulted from the 20th-century GR, such as land consolidation, proletarianization, and environmental degradation, can be avoided (Bumb 2009; Sanginga and Woomer 2009; Vanlauwe 2009). Through engaging in agroecosystems that integrate market innovations as well as modern agronomic techniques, farmers are envisaged as taking a market route to food security (Orr et al. 2001; Bationo 2004).

One of the central pillars of the AGR, institutional reform, encourages public private partnerships (PPPs). AGRA itself is an example of this, as an alliance between governments, international agricultural research institutions, private philanthropy and commercial interests. The market innovation element of the AGR includes novel credit mechanisms. Such methods act to drive the uptake of costly technologies such as fertilizer. The activities of two more PPPS illustrate how this works in the study context.

First, the Ghanaian Millennium Development Authority (MiDA) is one of the compacts funded by the USA's Millennium Challenge Corporation. MiDA provides fertilizer loans to smallholders with its Commercial Development of Farmer-Based Organizations project (International Fertilizer Development Center 2011). Through the Bank of Ghana, loans are made available to nongovernmental private finance initiatives, who then disburse them to organizations of 50 farmers. Second, the Ghana Grains Partnership involves the government of Ghana, international fertilizer producers and retailers Yara and Wienco, major Ghanaian and African banks, a broker, Prorustica, and the smallholder association Masara N'Arziki in the study site. The banks provide fertilizer credit to farmers through the cooperative (Guyver and MacCarthy 2011). In both MiDA and Masara N'Arziki schemes, farmers may repay credit at harvest time in maize or cash. By participating in these schemes, the Ghanaian government implements its commercial agriculture project, emphasizing the establishment of value chains and PPPS (Ministry of Food & Agriculture 2011). Governments, private companies, and smallholders thereby collude in the push toward a commercialized agroecosystem.

The promises of the AGR sound attractive. There is little question that most smallholders would appreciate higher yields and more profits. This article tackles the more important issue of what less beneficial elements the AGR may entail, and whether farmers need to risk getting involved in them to enjoy the benefits it offers.

COMMENTARY ON THE AGR

Debate over the AGR centers around land ownership and seed, issues that are inextricably linked to this article's central concerns of ISFM and marketization. Farmers use these resources to implement longstanding practices developed in response to their social as well as natural surroundings. This section, therefore, sets the discussion over the AGR in its socioeconomic and environmental context.

Land ownership is particularly contentious. In Northern Ghana, the dominant land tenure system reflects a pattern common across the continent. Chiefs have customary ownership of most land outside urban areas and allocate portions of this to individual families. Every generation, that land is subdivided among the sons of a patriarchal household head, and the resultant scarcity and continual cropping has been blamed for soil fertility decline. Farmers who operate under such customary systems do not hold legal title to the land they cultivate. Recognizing this, MiDAs land tenure facilitation project energetically supports the Ghanaian government's efforts at individual land titling that have been ongoing since the 1986 Land Title Registration Law. The argument underlying this effort is that land can act as loan collateral, and the capitalization this engenders facilitates more efficient production and, therefore, less risky market access (Deininger 1999). The evidence, however, does not always bear this out (Pinckney and Kimuyu 1994). Another dimension is added to the debate when the potential for "distress sales" of land is considered, resulting, in some cases, in land dispossession (Peters 2009; Jones-Casey and Knox 2011; Patel 2012). In the context of 21st-century land grabs by international companies, issues of land ownership have become even more important (High Level Panel of Experts on Food Security and Nutrition 2011; Obeng-Odoom 2013). In terms of the relationship between tenure and agricultural management, the dominant, but disputed, rhetoric is that farmers with legal titles are more likely to invest in their land, for example with organic ISFM or tree planting (Robertson and Pinstrup-Andersen 2010). The counter argument is that secure usufruct rights can be a sufficient incentive (Pretty et al. 2002; Food and Agriculture Organization of the United Nations 2001; Wiggins et al. 2011). When farmers are sure of secure continued access to land, for example, under the customary arrangements or family ownership systems they have traditionally engaged in, they may be equally willing to improve it as when they own formal titles.

Despite the importance of land, the less tangible input of labor has traditionally been conceived of as the limiting factor in West African agriculture (Swindell 1985). In the past, it derived largely from the family or from reciprocal group labor, sources that are difficult to cost. This is changing, however, as land shortage emerges and a labor market becomes more established. Farmers consider the relative availability of land and labor as they decide which of a range of land management techniques to implement, and the balance may make the difference between cultivating and fallowing a particular field. Some evidence for this can be seen in the way farmers fertilize their compound farms in West Africa: Preferential application of organic fertilizer to these easily accessible and relatively secure sites has been recorded throughout the 20th-century (Fortes et al. 1947; Prothero 1957; Harris and Yusuf 2001; Karbo and Agyare 2002).

A further consideration that influences farmers' practice is the availability of material inputs such as seed and agrochemicals. As farming systems become more commercialized these expensive inputs are becoming essential (de Ridder et al. 2004). One that has attracted particular attention in the debate over the AGR is seed.

The use of improved, hybrid seed, like land, is controversial. Despite giving better yield than landraces, its poor reproductive capacity and necessary annual purchase raise question about sovereignty and ownership: As farmers commercialize, they come to rely more on such external inputs over which they have less control (Amanor 2010; Thompson 2012). The suitability of exotic seed for maintaining ecosystem diversity is also questioned (Altieri 2009).

ISFM necessarily involves such external inputs, advocating the use of improved varieties alongside synthetic and natural soil fertility amendments. It has been defined as

> a set of soil fertility management practices that necessarily include the use of fertilizer, organic inputs, and improved germplasm combined with the knowledge on how to adapt these practices to local conditions, aiming at maximizing agronomic use efficiency of the applied nutrients. (Vanlauwe, Chianu, et al. 2010, 4)

Soil amendments such as inorganic and organic fertilizer are just part of a range of management techniques that include crop rotation, intercropping, mulching, and coppicing, and ISFM does emphasize the importance of using a variety of such methods. This stance arises in recognition of the complex decisions farmers make as they implement land management practices. They consider the availability and nature of physical and human resources as well as the demands of their preferred cropping pattern, and, therefore, use multiple SFM practices not only to maximize the benefits to the soil but also the effectiveness of the available production factors. Nevertheless, as the quotation above implies, within the AGR the emphasis is on amendments and inputs, and the issue of whether organics or inorganics should be prioritized has attracted much debate (e.g., Bationo et al. 2012). The thrust of the ISFM paradigm is that organic treatments are valuable primarily as a means toward increasing the agronomic efficiency of the "entry point," inorganic fertilizer (Vanlauwe, Bationo, et al. 2010).

In terms of agronomy, yield differences between organic and inorganic treatments are context specific and often negligible (Adediran et al. 2005; Jiang et al. 2006; Rasool et al. 2008; Ayuke et al. 2011; Pinitpaitoon et al. 2011), contributing toward the consensus that combined application of the two is optimum. The potentially directly deleterious ecological effects of inorganic fertilizer use, such as eutrophication and acidification, are less applicable in Northern Ghana considering the low rates applied to the sandy savanna soils. More relevant are the indirect effects of low soil organic matter reincorporation, resulting in lower soil water and nutrient retention. In the moisture and nutrient limited savanna these are primary considerations, and this environmental frame shapes farmers' agricultural practice. Indigenous technologies such as *zai* and *wafipa* compost pits have developed in response in East and West Africa respectively (Malley et al. 2001; Reij and Waters-Bayer 2001; Uphoff 2002).

More contentious is comparison of the relative costs of organics and inorganics and the resultant debate on subsidy. Poor production and distribution mechanisms and international trade tariffs mean fertilizers are at least four times as expensive in Africa as Europe (Druilhe and Barreiro-Hurle 2012), obstructing implementation of the Abuja declaration. Kofi Annan, as AGRA chairman, contextualizes this issue, noting that European and American farmers are heavily subsidized, in contrast to the conventional position on African policy (Wiggins and Leturque 2010; Annan 2012). Despite the yield increases that subsidies can engender (Denning et al. 2009), pro-free-market commentators opine that they should only be used to pave the way for market provision of unsubsidized fertilizer (Morris et al. 2007; Minot and Benson 2009). "Smart" subsidies like Malawi's voucher scheme, derided and lauded in equal measure, are the compromise solution (Dorward and Chirwa 2011). The emphasis on free-market fertilizer purchase is one facet of the AGRs valorization of a market-led ideology.

Marketization and profit generation are the core aims of the AGR. Some authors of an agroecological persuasion agree that local market access is an integral element of a functioning agroecosystem (Pretty 1995). Food sovereignty advocates are less convinced (Holt-Giménez and Altieri 2012). Market entry does increase exposure to risk. The issue is often getting enough goods to the market to make profit and getting them there on time, both of which involve financial outlay up front (Hazell and Poulton 2007; Oduro and Osei-Akoto 2007). Access to credit and infrastructural provision are, therefore, important in facilitating meaningful market engagement (Moseley et al. 2010). Another route to market involves assistance from an associate, such as an outgrower organization, that can bear some of the risk and facilitate links to inputs (Coulter et al. 1999; Robertson and Pinstrup-Andersen 2010). The question then becomes how much risk the respective partners absorb: the power relations between such organizations and the farmers they enroll are questioned by authors such as Porter and Phillips-Howard (1997). Diversification and subsistence continue to play risk reduction roles in such situations (Bryceson 1999).

The particular problematic of the AGR is thus nested in a matrix of environmental, social and economic concerns. Reflecting the complexity of the issues involved, there is some thoughtful commentary on the AGR that recognizes its potential advantages as well as its capacity to disempower farmers (Odame and Muange 2011; Sumberg et al. 2012). The challenge is to understand which elements of the AGR agenda may be beneficial and how they could be implemented so as to minimize disadvantageous effects. Simultaneously, some aspects may need to be treated with caution or rejected. In the complex and heterogeneous African agricultural environment, different combinations of solutions bear relevance between contexts, so the techniques of the AGR should form part of the "basket of options" (Ngwenya 2009) that livelihoods authors describe (Sanginga et al. 2009; Rakodi 2002). It is important to ascertain whether any element of the AGR in fact prevents farmers retaining the situation-appropriate strategies they have themselves developed over time to avoid risk.

The rest of this article tracks the progress of the AGR over five years in Dagbon, in the Guinea savanna of Northern Ghana. Here, about 25 miles from the regional capital of Tamale, farmers crop maize for subsistence and sale as part of a mixed farm-household system that incorporates grains, legumes, livestock, tubers, livestock, and, increasingly, vegetables. The savanna soils are fairly low in organic matter and nutrients (Kowal and Kassam 1978), and input and output market access has hitherto been quite limited (Bumb et al. 2011; Druilhe and Barreiro-Hurle 2012). Three linked studies are considered. Performed around the time AGR policies began filtering down to farmers, the first presents and explains farmers' opinions on the role of infrastructural provision for compost transportation. The second examines the relative advantages of organic and inorganic fertilizers and the third, five years on, reflects upon the effect of fertilizer credit provision. The data indicate that although the AGR advocates choice, this rhetoric is not always borne out in practice. The approach, when implemented as a package, is not, therefore, able to facilitate the livelihood choices that would benefit all farmers.

2008–2009: THE IMPORTANCE OF TRANSPORT INFRASTRUCTURE

The first in this set of investigations explored the factors that prevented Northern Ghanaian farmers from composting (see Bellwood-Howard 2013). In 2008, visits to communities in the Guinea Savanna revealed that, despite describing compost as more sustainable and advantageous, many semi-subsistence maize farmers were not using it extensively in their farms. To investigate what was constraining its use, a quantitative cross-sectional survey consulted 386 such farmers across three villages. The farmers were pursuing the rained subsistence maize farming typical of the area, with groundnuts as additional cash crops. Many also kept cattle, small ruminants, and poultry. Sampling was purposive and stratified. In two villages, Zaazi, in Savelugu-Nanton district, and Ypilgu, in Tolon-Kumbungu district of Ghana's Northern Region, the male household head and, if available, one of his wives were interviewed using a structured questionnaire. These were complemented by interviews with one junior man and woman from every household (there were no female headed households, and one household comprised only bachelors). In the third village, Yong, in Savelugu-Nanton district, composting was less common, so one male and one female respondent were interviewed from every household. The sample size thus exceeded 10% of the combined population of all three communities. When a target respondent was unavailable, for example due to migration, an alternative individual of similar status was sought from within the same household. The main question was "What do you find the most difficult aspect of composting?" Respondents were able to choose from 17 categories that had been derived from content analysis of 30 in-depth qualitative semi-structured interviews with residents of the same villages. They identified the factors they considered the most, the second most and the third most constraining. Demographic data was also collected for each of the respondents. The descriptive statistics presented here were produced using SPSS (Ver. 16). As in all the three studies, the ethnographic methods of participant observation and informal conversation were indispensable. As no compost was being made in the 2008–2009 dry season when this survey took place, such participant observation happened in the course of daily household life.

The intention had been to weight the responses to the main question according to their ranking as first, second, or third worst. However, Figure 1 shows that farmers were overwhelmingly of the opinion that poor transportation facilities were the primary barrier to composting. Even when weighting was applied, this result held, so the raw data for the "worst" barrier are presented.

This result was interpreted as indicating a lack of financial capital: when asked to suggest solutions to this problem, 47.1% of the whole sample suggested replacing the small, cheap bicycles and headpans they had hitherto been using with larger, more expensive vehicles (*benzirra*), like hand-pulled trucks, wheelbarrows, or donkey carts.

In 2012, return visits to these communities in farming season showed that in Ypilgu, farmers had begun to implement such a solution. Individuals

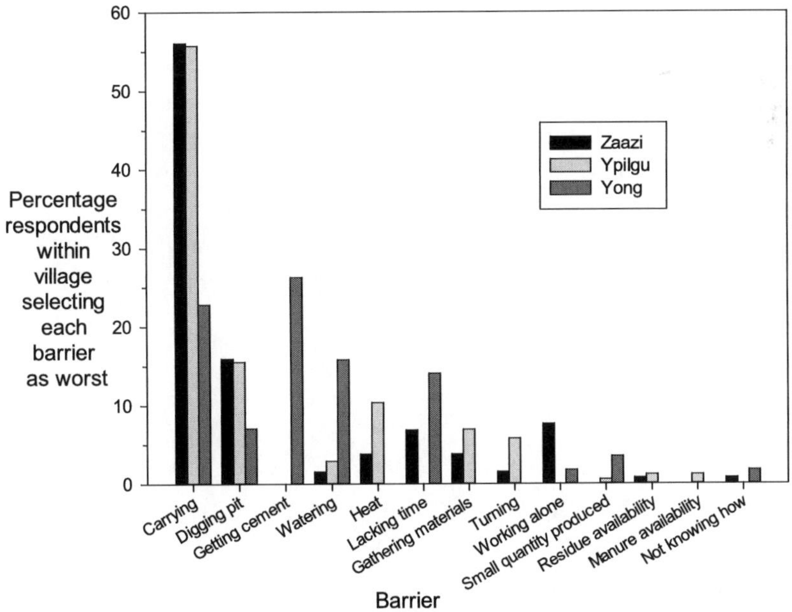

FIGURE 1 Proportion of respondents in Zaazi, Ypilgu, and Yong villages selecting each barrier to composting as worst.

had begun to purchase bullock carts, to the extent that five farmers owned them by 2013 and a market for their hire had emerged. Individuals' personal financial capital accumulation facilitates ownership or rental of such *benzirra*, so this qualitative data seems to indicate that local market engagement could conceivably go some way toward solving this transport problem and promoting more sustainable fertilization practice based on organics. The actions of the farmers who participated in the 2010 and 2011 research also indicated that traditional mechanisms of reciprocity and gift giving would allow those who owned vehicles to share them with other community members (Bellwood-Howard 2012). However, issues of equity emerge: This option is, of course, more available and more advantageous to richer farmers, reminiscent of the economic differentiation that resulted in some instances immediately after the first Asian Green Revolution (e.g., Dhanagare 1987; Jayatilaka 1989; Webster 1990; Harriss-White and Janakarajan 1997).

The state also has a role to play in such a situation: for reasons of scale it is usually held responsible for the infrastructural improvements essential for well-functioning transport systems (Morris et al. 2007). In Yong, the government, in the spirit of the AGR, brought in private Chinese contractors to grade and tarmac the road to the local town. In Ypilgu, however, the local authority had delayed for five years in making improvements to the feeder road, so motorized transport came to the village less than once a month. This was less of a constraint on the operation of the bullock hire market than on farmers' access to inorganic fertilizer. High transport costs are one major reason for high fertilizer prices on the continent (Jayne et al. 2003) and the Ghanaian government, like others, has responded with the fast and visible solution of fertilizer subsidy, alongside the ongoing development of transport infrastructure. There have also been suggestions that such a structural solution as subsidy could be used to overcome, for example, the high import costs of cart axles (Reddy 1988). Agroecologists and technologists, therefore, both acknowledge that transportation and import tariff infrastructure is essential for environmentally as well as economically sustainable SFM (see, e.g., Toenniessen et al. 2008; but also Altieri 2009). Public investment, thus, has a role to play as well as the private capital that can provide ownership and hire of the vehicles that are used on public roads.

Implications of the study: State and private infrastructural investments are necessary for sustainable SFM.

2010: INTEGRATED SOIL FERTILITY MANAGEMENT AND COMPOST

In 2010, a further study compared the effects of compost and fertilizer application on maize performance. Twenty-nine men and one woman participated across two villages, Ypilgu and Zaazi. Each allocated an area

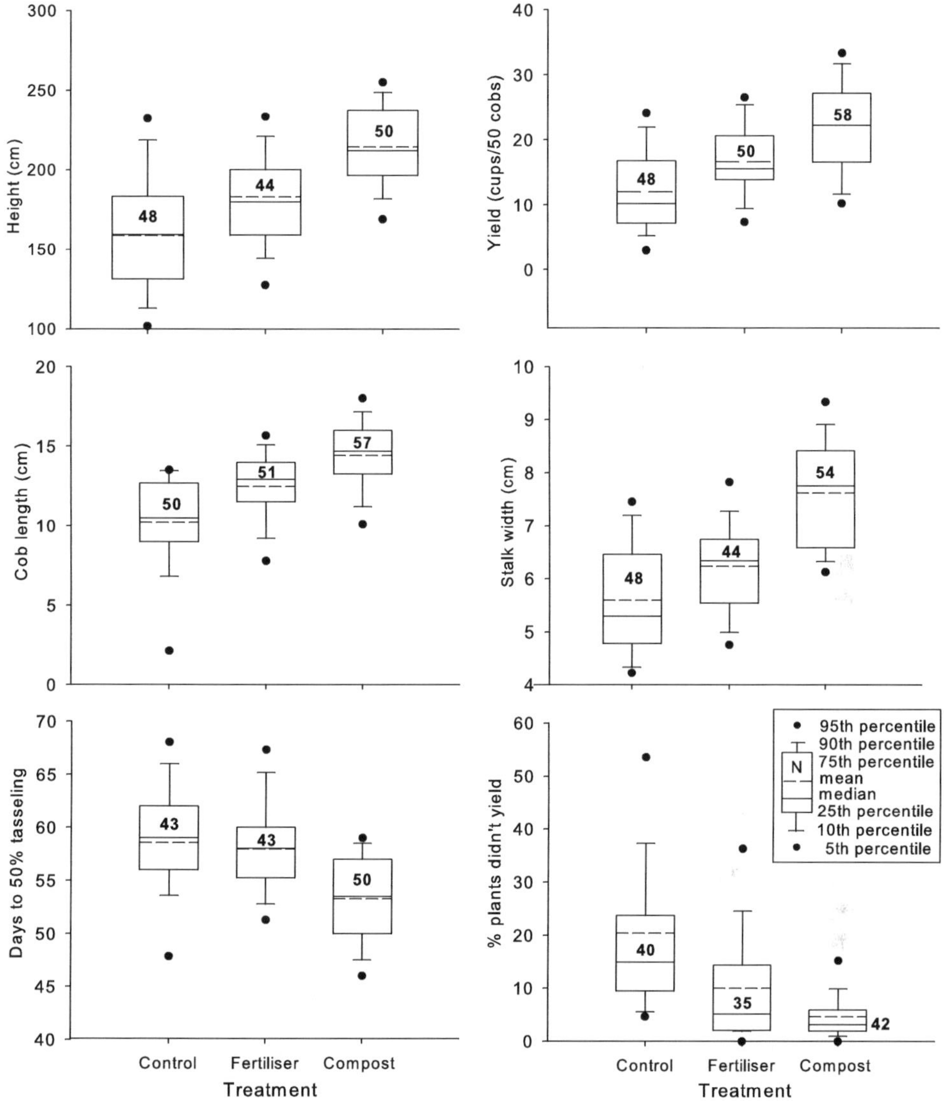

FIGURE 2 Selected growth parameters and yield of maize plants in control, inorganically fertilized and composted plots (*N* = sample size).

within one of their fields to be fertilized only with compost, an area to be fertilized only with inorganic fertilizer and a control area that received no organic or inorganic amendments in the experimental year. Farmers used these inputs at the rates that they were able to manage given their resource endowments. Maize was planted on the plots and the yield and growth parameters indicated in Figure 2 were measured. Soil was tested for C, N, P, K, and pH. The experiment did not use a combined treatment, because the aim was to ascertain whether farmers with a limited capital endowment could use compost or inorganic fertilizer most effectively within the larger matrix of their SFM practice, and hence where the emphasis should lie. A further aim was to explore the mechanisms by which the fertilizer and compost influenced maize growth. Of course, the effects of the multiplicity of SFM methods that farmers use do endure over more than a single cropping season, so the experiment could not be said to be controlled, as control plots were likely to have retained residual fertility from previous years. In addition, as different farmers' fields were used, factors such as weed control regime varied widely. These conditions were tolerated because the experiment aimed to relate to farmers' real resource allocations, and this was held to be a more valid concern than which treatment performed better in a controlled experiment. SPSS was used to perform Mann Whitney U tests on the data.

The results were unequivocal: At farmers' application rates, maize plants treated with compost yielded better and also outperformed those treated with inorganic fertilizer for five other growth parameters.

Soil nutrient tests showed that this was unlikely to be as a result of better macronutrient provision. Mann-Whitney U tests on the data shown in Table 1 showed that there were no significant differences in macronutrient levels between the compost and inorganic fertilizer treatments, although pH was higher for the compost treatment at tasseling.

This similarity between the nutrient levels provided by compost and inorganic fertilizer resonates with results from the literature. There is no universal trend; whether organic or inorganic amendments give higher nutrient levels is context dependent. For example, Ouedrageou et al. (2006) found that, in Burkina Faso, the extent of tillage determined whether compost or inorganic urea resulted in higher soil nitrogen. For Eghball and Power (1999) in the temperate context of mid-western America, application of ammonia fertilizer gave higher soil N levels than application of compost, but only in drier years. The importance of such context-specific non-nutrient factors serves to emphasize that, in the heterogeneous African environment, appropriate management techniques vary between sites. In the study sites, this was confirmed by qualitative data collection.

In the 2010 farming season, farmers were interviewed in their fields once before and once after the experiment. Participant observation also took

TABLE 1 Nutrient concentrations in compost and fertilizer amended plots at tasseling stage

Parameter	Fertilizer	Compost	Mann-Whitney *p*
C (%)	1.12	1.13	0.804
N (%)	0.11	0.13	0.823
P (mg/kg)	9.73	9.15	0.354
K (mg/kg)	135.23	128.64	0.725
pH	7.71	6.76	<0.001

place during compost preparation, transportation, and application. These data indicated that the superior performance of organic amendments in this setting resulted from their ability to increase soil water retention capacity. In interviews, as well as informal conversations, farmers described how compost improved soil structure. Four farmers demonstrated how they could recognize the high water retention capacity of soil amended with compost when their foot sank into it, in contrast to the hard, compact soil treated with inorganics. Maize yields can be affected after 5–15 dry days (Fischer et al. 1982; Eck 1986; Barron et al. 2003; Recep 2004), and farmers in the study site agreed that the enhanced ability of compost-amended soil to retain moisture became beneficial to maize after a week without rain.

A quantitative assay was, therefore, performed to test this property. In Ypilgu village, thirty infield compound farm sites fairly high in organic matter and thirty outfields or "bush farms" with lower organic content were selected, using Munsell color as a proxy for organic matter (see Wopereis et al. 2006). Soil water content by weight was measured immediately after a heavy rainfall event and then 6 and 11 dry days later. To avoid destroying soil structure by removing samples from the field, the assay was performed in situ. It was, therefore, necessarily carried out in March, before the rainy season began in earnest, so that dry spells between rainfall events were likely to last a week or more. The practice of preferential organic matter application to compound farms pertains in the study sites (Karbo and Agyare 2002), and in March farmers had not yet carried compost to their bush farms, hence why compound infields and bush farm outfields were used as proxies for sites with and without organic matter amendments. The farmers' contention that fields with historic compost application retained more water was borne out: infield soils as a group retained significantly more water both initially and after 11 days (see Figure 3).

The main advantage of compost in this site, its ability to enhance water retention capacity, is a function that inorganic fertilizers cannot perform. This resonates with Ekins et al.'s (2003) concept of strong sustainability: some functions of natural capital cannot be substituted for (Neumayer 2003). As fertilizer cannot serve moisture retention functions, indications are that within an ISFM strategy for the savanna, the emphasis should rest upon organics. This consideration resonates with Sanchez's (1994) "second paradigm" (68), the forerunner to ISFM, which advises farmers to rely more on biological processes by adapting germplasm to adverse soil conditions, enhancing soil biological activity, and optimizing nutrient cycling to minimize external inputs and maximize the efficiency of their use.

It is important to contextualize these results within the matrix of ISFM practices and their associated resource requirements. The field experiment was limited to testing amendments, yet any one of the techniques that raise soil water retention capacity could play a valuable role here. Farmers in the

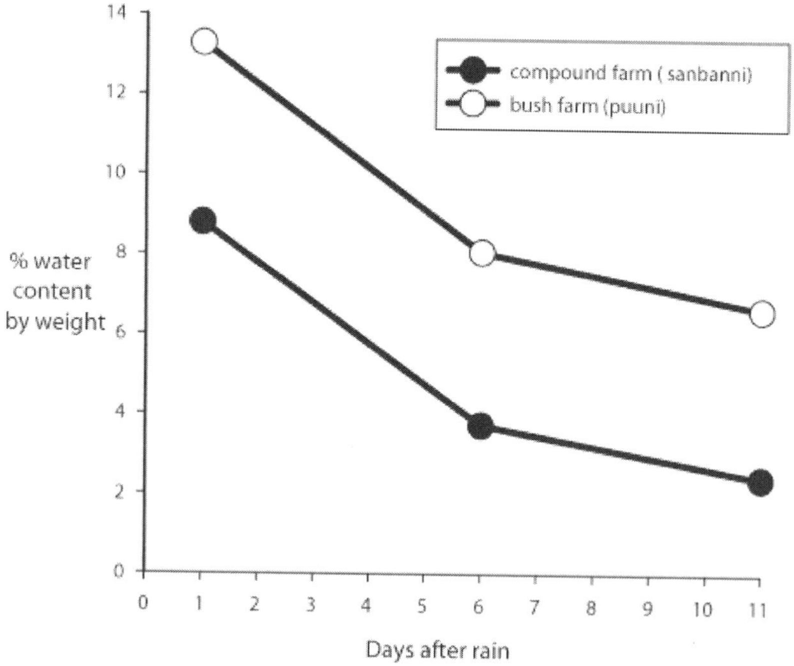

FIGURE 3 Mean water content of 30 compound and 30 bush farms up to 11 days after rain. Differences between compound and bush farms were significant (Mann-Whitney $p < 0.001$) at each sampling date.

study sites did practice such strategies, including annually rotating the portion of the field that compost was applied to, incorporating stover and double ploughing with bullocks. The last method involves ploughing once, allowing grasses to grow back, then four weeks later using bullocks to plough those grasses 15 cm into the topsoil, rather than 30 cm deep as a tractor would do. Farmers' use of such strategies shows that over time they have developed SFM methods that meet their needs for organic amendments. Furthermore, whereas organic treatments have high labor costs, fertilizers are expensive. For most farmers in this site, the cash required for fertilizer application had hitherto been more limiting than the labor costs of composting, and this plays a role when famers evaluate the relative costs and advantages of each technique.

However, although organic amendments are generally the most appropriate soil treatment in the study site, this cannot be taken as an indication that the same is true all across the region. Some farmers were observed performing mixed application, the strategy advocated within ISFM. In interview, one farmer from Ypilgu explained that this would be his preferred practice, as it would improve water retention capacity as well as raising absolute nutrient content to superlative levels. Nevertheless, limited availability of both organic and inorganic fertilizers constrained him from applying this treatment to every field. When the realities of variable resource availability meant farmers had to choose between the two, organics were preferred. This study did not implement a combined treatment that would facilitate a conclusion about the agronomic efficacy of mixed application, as the aim was rather to observe how farmers' ISFM decisions interacted with their ability to use different resources. However, it is those same issues of resource accessibility that imply a role for inorganic fertilizer within a regional SFM strategy. The importance of site specificity has been emphasized: Different soils do respond to organic and inorganic amendments in various ways, and the availability and transportation of organic materials can be more or less problematic between locations (Bationo and Mokwunye 1991b). At the moment labor seems less limiting than cash for these particular farmers. However, before the introduction of bullock carts, compost transportation was, indeed, a challenge in the study site. Fertilizer is, therefore, potentially useful in situations where compost may be less accessible or of poorer quality than that applied by the farmers in these villages.

The real value of these results is to illustrate that ISFM is most usefully interpreted not only as mixed application of organics and inorganics, as the AGR emphasizes, but the ability to apply whichever combination thereof suits the resource availability and crop demand context. This could include sole organic addition, as seemed the best choice here.

Fertilizer subsidy may, therefore, be helpful in the contexts where inorganic amendments have a role to play. However, considering these results, it is worth considering whether the funds allocated to mineral fertilizer subsidy could usefully be diverted toward promoting development of organic sources, for example by subsidizing their transportation.

Reflecting on the type of germplasm that farmers use, it should also be noted that the experiment was performed on the high-yielding improved 'white' maize varieties that most farmers in the villages have adopted. These are the strains that are shown performing best under compost in Figure 2.

Implications of the study: ISFM should be interpreted as a range of possible practices, which may include sole organic application. Organics should be emphasized for their soil water retention properties.

2012: INPUT MARKET ACCESS AND CREDIT

The final study in this trio examined the impact of fertilizer credit schemes, including the MiDA scheme and its government and informal parallels, on farmers' SFM and commercialization strategies. It compared socioeconomic characteristics and fertilization practices of 104 farmers who had borrowed fertilizer credit and 101 who had not, across five villages: Ypilgu, Tarakpaa,

Zugu, Satani, and Yirikpani. Credit group members were identified from lists provided by the chairman of their farming group and invited to participate in a structured survey with a qualitative component. They then self-selected, and those who agreed to participate were interviewed until the sample size reached 5% of the population of the settlements. Non-credit borrowers who matched them were then selected by a method similar to the nearest neighbor technique (Kennedy and Cogill 1987), in this instance matching farmers on the basis of their household size, their approximate age and their position within their household. These farmers guessed their ages to be from about 16 to about 70 years old. Only two were women. Descriptive and inferential statistics were performed on the data using SPSS.

The results of the survey showed that credit group members generally applied more fertilizer, achieved higher yields and sold more maize. Crucially, they were also more likely to use compost (see Table 2).

Across Africa, richer farmers often do use more compost than the less well off (Vanlauwe and Zingore 2011). Indeed, there were six wealthier farmers who illustrated this. They had large herds of cattle that fertilized their fields. They could afford all the fertilizer they needed, and were engaging in markets to some extent: one even owned a tractor with which he drove a mechanized sheller, processing his neighbors' grain, collecting a tenth of their harvest as payment and selling some of it when the price peaked. However, these farmers had not joined credit groups, because they didn't need to: the son of one explained that he considered entering the market on 'your own strength' preferable. To some extent these results bolster the AGR rhetoric: These farmers have exceeded the need for credit mechanisms and are also able to pursue more sustainable SFM. If credit were to facilitate such wealth for their peers, they could also ostensibly attain such an achievement. Two factors, however, disrupt this reading.

TABLE 2 Differences between credit group members and non-members

Parameter	Group members' mean	Non-members' mean	Mann Whitney *p*
Number of 50 kg bags of fertilizer used	5.90	4.41	0.000***
Fertilizer application rate (bag/acre)	1.61	1.41	0.167
Bags of fertilizer bought	2.06	4.02	0.000***
Bags of fertilizer borrowed from sources other than the credit group	0.11	0.36	0.333
Yield (50 kg bags maize/acre)	5.05	4.79	0.316
Yield per household member	0.37	0.31	0.028*
Bags of maize sold	1.70	1.05	0.000***
Percentage of maize sold by bag	8.37	7.89	0.082
Income received from maize sales (GhC)	124.42	91.62	0.021*
Income per bag of maize	52.3	51.3	0.794
Percentage of farmers using compost	34	27	

***=significant at 0.000 level; *=significant at 0.05 level.

The first explains the association between credit group membership and compost application. A disproportionate number of household heads were credit group members, and a higher percentage of them than of the younger men used compost: 41.3% as opposed to 17.7% (X^2 $p < 0.001$). This relationship was more statistically significant than that between group ownership and compost application, and has more important implications. The fact that household heads were more likely to use compost and, on average, used more of it, points to an aspect of agriculture the AGR deals less well with. It is the traditional responsibility of the Dagomba household head to feed his family, and several respondents described the pride they felt if able to use their own maize do so for much of the year. This could be seen as expressing the cultural meaning of subsistence, but risk avoidance is another contributing factor. Maize is more nutrient-hungry and water responsive than the groundnuts that young men traditionally cultivate for cash, and household heads, generally better endowed with cattle and with access to more labor, are often better able to supply manure and compost to this cereal crop. This is why subsistence maize is grown in compound farms, to which the landlord organizes the application of compost. Their responsibility to feed their dependents also drives their uptake of fertilizer credit, and the confluence of these two household food security mechanisms is responsible for the association of credit and compost use. The current generation of household heads has not replaced organic-subsistence maize cultivation with practices that represent a market route to food security. Whether the next generation will do so remains to be seen, but farmers' responses to the question of whether maize was a commercial or subsistence crop, shown in Table 3, give some indication.

Farmers' preference for organic, subsistence maize cultivation is a response to their specific context of agroecological risk. It is, therefore, a knowledge resource that strategies like the AGR could usefully learn from rather than undermining.

The second factor that sits uneasily with the AGRs promotion of credit is the complaint of a minority of farmers about the schemes examined here.

TABLE 3 Farmers' responses to the question of whether maize is a commercial or a subsistence crop

Farmer response	Percentage of 205 farmers giving this response
They would sell maize only if some remained at the end of the year	70.2
Maize can be sold as well as eaten	52.9
Maize is a subsistence crop	3.4
Maize is a commercial crop	0.5

Note. Many farmers gave multiple responses to this open question. As answers in the first two rows are not mutually exclusive the total exceeds 100%.

Some saw the terms of access as unfavorable, to the extent that in 2012 a group in Tarakpaa declined participation in the MiDA-funded group they had formerly been part of. They were aware that by repaying fertilizer loans in maize at harvest time they received the minimum price for their produce. This meant that in effect they paid up to twice as much for fertilizer as they would if they had bought it with cash. Farmers generally borrowed one 50kg bag of compound fertilizer and one of ammonium sulphate top-dresser per acre, worth subsidized prices of 30 and 25 GhC, respectively, in 2011 (1 GhC = approximately 0.6 USD in 2011). Repayment rates were 1–1.5 bags of maize per bag of compound fertilizer and 1 bag of maize per bag of top dresser. At harvest time, a bag of maize cost around 30 GhC. Six months later it could sell for up to 60 GhC. Both NGOs and private lenders took advantage of this to cover costs or make profit. Twenty five percent of the farmers who had opted out of the MiDA-funded scheme explained that they would only join another group in future if it offered them credit on fairer terms. Forty eight of the 104 group members, when asked, made negative comments about what they perceived to be the inequitable effects of the credit schemes. These are listed in Table 4.

Nevertheless, despite the opinions of these particular farmers, the majority appreciated the help (*suŋsim*) fertilizer credit afforded them: 94 of the 104 group members sampled said that, despite their few concerns, group membership was advantageous overall.

For most Dagomba farmers, both household heads and young men, some access to input as well as output markets is better than none. The

TABLE 4 Farmers' concerns about fertilizer credit schemes

Statement	Number of farmers	Percentage of entire sample	Percentage of group members
Concerns related to credit provision			
Inputs are delivered late	19	9.3	18.3
Repayments are expensive	10	4.9	9.6
They prefer self-sufficiency	10	4.9	9.6
They would be unable to repay in the event of drought and crop failure	10	4.9	9.6
MiDA profits more than farmers	5	2.4	4.8
Some have not harvested when repayments are due	3	1.5	2.9
Market prices are low when repayments are due	3	1.5	2.9
Concerns related to fertilizer			
It is problematic in case of drought	55	27	
There is low soil fertility in years following application	24	12	
Weeds thrive in years following application	10	5	
It has negative health effects	7	3	

AGR rhetoric promotes the linked input-output market mechanisms of contract farming and outgrowing over government and informal credit routes, and the problem of a lack of sanctions for non-repayment in government schemes certainly was observed in this case. Credit does, however, need to offer access at minimal cost and risk. If commercial schemes increase risk and make farmers worse off, they are a high price to pay for fertilizer that, as has been seen, is in any case a second best option after organic matter in this particular situation. Indeed, none of the farmers in this sample were members of outgrower credit initiatives, even though two of the communities were in district where a Ghana Grains Partner, the Masara N'Arziki association, was operating. Some did not have the five acres required to join that organization. However, larger scale farmers who were asked in the qualitative section of the survey about their reasons for not joining the cooperative cited discomfort with the requirement that all maize be sold back to the creditor. The high fertilizer levels Masara N'Arziki asked outgrowers to apply to their fields were also problematic for some, with 91 farmers mentioning at least one of the concerns relating to fertilizer use in general listed in Table 4. In sum, the farmers in this sample preferred to satisfy their subsistence needs using largely non-marketized transactions and organic soil amendments.

Implications of the study: Credit and market access are only helpful when they are optional, low risk and not tied to conditions.

DISCUSSION

This article has seen that the AGR incorporates some appropriate elements of farmers' practice. Yet, it combines them with other, less advantageous schemes and, despite the rhetoric of choice and site specificity, applies them as a package in a hegemonic fashion that means the approach as a whole does not meet all farmers' needs.

With respect to ISFM, organic matter was shown to be of paramount importance in this particular location because of its ability to enhance water retention capacity, indicating the validity of Sanchez's (1994) second paradigm. This does not mean that fertilizer is always irrelevant: There could be a case in some areas for mixed application, a treatment that was not tested here but has been widely endorsed in the literature. However, as well as the agronomic advantages of each treatment, it is essential to consider the availability of the resources necessary to apply them. The financial cost of fertilizer as well as the fact that it does not provide a soil water retention function means that in this site, organic amendments comprise a more useful resource. This is a good explanation for why richer farmers chose to invest in bullock carts that facilitated a market in compost transport, rather than purchasing fertilizer. What is most significant about this result is that it demonstrates the importance of context specific solutions. Indeed, the importance of organic matter additions to the soil in these Northern Ghanaian villages does not imply that compost is always necessary here: A farmer may equally implement any other strategy compatible with their resource capabilities that integrates organic matter into the soil, such as double ploughing. Similarly, sole fertilizer application could be the best solution in other sites, and combined application has been shown to be most effective in many others (Bationo and Mokwunye 1991b; Chivenge et al. 2011; Tittonell et al. 2011). Although the AGR nominally recognizes that various SFM systems are appropriate for different sites, it largely interprets ISFM as combined application. It does not support the idea that that organic amendments or agroecological techniques promoting generation of soil organic matter could potentially be sufficient in sole application, as has been shown to be the case in these particular villages. This results from an emphasis on nutrient rather than soil water provision (Bationo and Mokwunye 1991a). This is not necessarily universally relevant, especially in terms of the resources farmers are able to access. The AGR, therefore, struggles to reflect the holistic approach that farmers develop through a working familiarity with their agroecosystem and knowledge of available resources. Its claim to recognize the importance of context specificity is thus not borne out, because it deals less well with sites requiring strongly sustainable characteristics like soil water availability that cannot be purchased as an input.

The same idea of choice is applicable to market strategies. Farmers use a range, involving commercialization but also subsistence and reciprocity, as a hedge against market as well as agroecological risk. Credit can, therefore, be useful as it can facilitate market access if and when farmers need it. However, absolute market integration, of the type that Masara N'Arziki advocates, is just as risky as none. In addition, harvest-payment credit mechanisms like the one encountered in this study are a sure route to inequality. In contrast to the idea that the AGR gives farmers access to multiple livelihood options, they lock farmers into arrangements that, ironically, disadvantage them in the marketplace, because they are designed to make profit for another agency such as an NGO or outgrower association.

The systems and mechanisms that farmers have developed autonomously, such as organic matter incorporation and subsistence farming, are generally attuned to their own agroecological and market conditions. The most useful function the AGR could perform would be to provide low risk entry routes to techniques that farmers do want but have hitherto been unable to access. The problem is that, despite advocating farmer choice, it deals less well with site-specific demands that fall outside its overall objectives of marketization and combined application. Farmers are encouraged to access the market in any way they like, as long as they would like to access the market. Likewise, they are encouraged to practice any sort of combined application of inorganic and organic amendments they like, as long as they practice combined application. As long as the options of the

AGR are presented to farmers as part of such a package, they are beneficial to only a section of the population.

This is especially the case if those options themselves run the risk of accentuating socioeconomic differentiation. This is the case with almost any change to agrarian systems: the bullock cart hire market that farmer themselves instigated did so. Land titling is similar: in the 20th-century GR it resulted in some land consolidation (Niazi 2004). The evidence presented here indicates that some such innovations may be of dubious utility and it is, therefore, questionable whether farmers need to take the risk of implementing them. Having a title to land, for example, is not a prerequisite to investing in it: farmers in the study situation do not have legal titles to their land but are already using strategies to enhance its soil organic matter content. Similarly, they are already able to access credit without using land as collateral. Warrantage systems, where farmers are able to store produce in warehouses in order to wait for a higher price (Tabo et al. 2011) pose another, potentially less risky, alternative to the types of credit schemes described in this study. In warrantage, the produce acts as collateral, making land titles unnecessary. Warrantage has been implemented on a small scale in Ghana (e.g., Ghana News Agency 2010) and a warehouse receipt system began in Northern Ghana in 2013 (ACDI/VOCA 2013). Such initiatives seems to hold more potential than land titling exercises for rural areas like the study area, where land is owned by chiefs in trust for families.

These studies indicate that basic infrastructural support would do more than targeted input schemes to enable farmers to perform the management strategies that they have developed over time, such as compost application, and take advantage of new opportunities that present themselves. Warrantage, for example, requires warehouse facilities. Such investments point to the involvement of large organizations like government, with extensive capacity to provide background services and less imperative to make profit than private sector bodies.

Almost all commentators on the AGR nominally agree that single solutions cannot work across the smallholder spectrum. However, whereas the movement's advocates consider that the multiple pathways available within the package are sufficient to meet the needs of all, critics are of the opinion that non-commercial alternatives must be retained, so that farmers in different sites may choose subsistence, organic and landrace as well as market, ISFM and improved variety options (Amanor 2010). The farmers encountered in these studies preferred inputs and markets over which they had more control, which for them meant organic amendments and subsistence mechanisms. These must, therefore, be conserved, and external tools like subsidy can potentially support such activity. There is little need for farmers to risk getting involved in mechanisms such as unfair credit, fertilizer application and land titling when they already have developed the agroecological strategies they need to lead sustainable agricultural livelihoods.

REFERENCES

ACDI/VOCA. 2013. *First grain warehouse receipts issued in Ghana*. http://www.acdivoca.org/site/ID/news-First-Grain-Warehouse-Receipts-Issued-Ghana (accessed September 19, 2013).

Adediran, J., L. Taiwo, M. Akande, R. Sobulo, and O. Idowu. 2005. Application of organic and inorganic fertilizer for sustainable maize and cowpea yields in Nigeria. *Journal of Plant Nutrition* 27:1163–1181. doi:10.1081/pln-120038542

African Union. 2006. *Africa Fertilizer Summit—Abuja declaration on fertilizer for the African Green Revolution*. Abuja, Nigeria: African Union.

Altieri, M. 2009. Agroecology, small farms, and food sovereignty. *Monthly Review* 9:102–113.

Amanor, K. 2010. *Participation, commercialization and actor networks. The political economy of cereal seed production systems in Ghana*. Working Paper 16. Brighton, UK: Future Agricultures Consortium.

Annan, K. A. 2012. Achieving food and nutrition security for all [Keynote address]. In *Securing Food. Harvesting the Future! Existing and New Approaches to Food Security—A contribution of German Development Policy*. Berlin, December, Berlin.

Ayuke, F., L. Brussaard, B. Vanlauwe, J. Six, D. Lelei, C. Kibunja, and M.Pulleman. 2011. Soil fertility management. Impacts on soil macrofauna, soil aggregation and soil organic matter allocation. *Applied Soil Ecology* 48:53–62. doi:10.1016/j.apsoil.2011.02.001

Barron, J., J. Rockstrom, F. Gichuki, and N. Hatibu. 2003. Dry spell analysis and maize yields for two semi-arid locations in East Africa. *Agricultural and Forest Meteorology* 117:23–37.

Bationo, A., Ed. 2004. *Managing nutrient cycles to sustain soil fertility in Sub Saharan Africa*. Cali, Colombia: CIAT.

Bationo, A., T. Fairhurst, K. E. Giller, V. Kelly, A. Mando, P. Mapfumo, et al. 2012. *ASHC handbook for ISFM*. Wallingford, UK: CAB.

Bationo, A., and Mokwunye, A. 1991a. Alleviating soil fertility constraints to increased crop production in West Africa. The experience in the Sahel. *Nutrient Cycling in Agroecosystems* 29:95–115. doi:10.1007/BF01048992

Bationo, A., and Mokwunye, A. 1991b. Role of manures and crop residue in alleviating soil fertility constraints to crop production. With special reference to the Sahelian and Sudanian zones of West Africa. *Nutrient Cycling in Agroecosystems* 29:117–125. doi:10.1007/BF01048993

Bellwood-Howard, I. 2012. Donkeys and bicycles. Capital interactions facilitating timely compost application in Northern Ghana. *International Journal of Agricultural Sustainability* 10:315–327. doi:10.1080/14735903.2012.666030

Bellwood-Howard, I. 2013. Compost adoption in Northern Ghana. *International Journal of Sustainable Development & World Policy* 2(2):15–32.

Bryceson, D. 1999. *Sub Saharan Africa betwixt and between. Rural livelihood practices and policies*. DeAgrarianisation and Rural Employment Network Working Paper 43. Leiden, the Netherlands: Afrika StudieCentrum.

Bumb, B. 2009. *Issues and challenges in fertilizer market development. Challenges in developing agricultural input markets in Africa*. Africa Agricultural Markets

Programme Fertilizer Symposium "Getting Fertilizers to Farmers: How to do it? Who should do it? and Why it Should be Done?" Livingstone, Zambia, June 15–1, Common Market for Eastern and Southern Africa, World Bank.

Bumb, B., M. Johnson, and P. Fuentes. 2011. *Policy options for improving regional fertilizer markets in West Africa*. IFPRI Discussion Paper 01084. Washington, DC: IFPRI.

Chivenge, P., B. Vanlauwe, and J. Six. 2011. Does the combined application of organic and mineral nutrient sources influence maize productivity? A meta-analysis. *Plant Soil* 342(1–2):1–30. doi:10.1007/s11104-010-0626-5

Coulter, J., A. Goodland, Tallontire, and R. Stringfellow. 1999. Marrying farmer cooperation and contract farming for service provision in a liberalising Sub-Saharan Africa. *ODI Natural Resource Perspectives* 48:1–4.

Deininger, K. 1999. Making negotiated land reform work: Initial experience from Colombia, Brazil and South Africa. *World Development* 27:651–672. doi:10.1016/S0305-750X(99)00023-6

Denning, G., P. Kabambe, P. Sanchez, A. Malik, R. Flor, R. Harawa, et al. 2009. Input subsidies to improve smallholder maize productivity in Malawi. Toward an African Green Revolution. *PLoS Biol* 7(1):e1000023. doi:10.1371/journal.pbio.1000023.

de Ridder, N., H. Breman, H. van Keulen, and T. Stomph. 2004. Revisiting a 'cure against land hunger'. Soil fertility management and farming systems dynamics in the West African Sahel. *Agricultural Systems* 80:109–131.

Dhanagare, D. 1987. Green Revolution and social inequalities in rural India. *Economic and Political Weekly* 22(19–21):AN137–AN144

Dorward, A., and E. Chirwa, E. 2011. The Malawi agricultural input subsidy programme. 2005-6 to 2008-9. *International Journal of Agricultural Sustainability* 9:232–247.

Druilhe, Z., and J. Barreiro-Hurle. 2012. *Fertilizer subsidies in sub-Saharan Africa*. ESA Working Paper No. 12-04. Rome: Agricultural Development Economics Division: FAO.

Eck, H. 1986. Effects of water deficits on yield, yield components and water use efficiency of irrigated corn. *Agronomy Journal* 78:1035–1040.

Eghball, B., and F. Power. 1999: *Phosphorus- and nitrogen-based manure and compost applications corn production and soil phosphorus*. Paper No. 12133. Lincoln, NE: USDA-ARS and University of Nebraska Agricultural Research Division.

Ekins, P., S. Simon, L. Deutsch, C. Folke, and R. de Groot. 2003. A framework for the practical application of the concepts of critical natural capital and strong sustainability. *Ecological Economics* 44:165–185. doi:10.1016/s0921-8009(02)00272-0.

Fischer, K., E. Johnson, and G. Edmeades. 1982. *Breeding and selection for drought resistance in tropical maize*. In Symposium on Principles and Methods in Crop Management for Drought Resistance with Emphasis on Rice. Philippines: International Rice Research Institute.

Food and Agriculture Organization of the United Nations. 2001. *The economics of soil productivity in Sub-Saharan Africa*. Rome: FAO.

Fortes, M., R. Steel, and P. Ady. 1947. Ashanti Survey 1945—46. An Experiment in Social Research. *The Geographical Journal* 110:149–170.

Ghana News Agency. 2010. *NGOs provide food security for 3,334 households in Wa East.* Available at http://www.modernghana.com/news/261851/1/ngos-providefood-security-for-3334-households-in-.html. (accessed on July 6, 2012).

Guyver, P., and M. MacCarthy. 2011. The Ghana grains partnership. *International Journal of Agricultural Sustainability* 9:35–41. doi:10.3763/ijas.2010.0564

Harris, F., and M. A. Yusuf. 2001. Manure management by smallholder farmers in the Kano close settled zone, Nigeria. *Experimental Agriculture* 37:319–332. doi:10.1017/S0014479701003040

Harriss-White, B., and S. Janakarajan. 1997. From Green Revolution to Rural Industrial Revolution in South India. *Economic and Political Weekly* 32:1469–1477.

Hazell, P., and C. Poulton. 2007. *Experiences with commercial agriculture. Case study on food staples. All-Africa review of competitive commercial agriculture in Sub-Saharan Africa study.* Washington, DC: FAO-World Bank.

High Level Panel of Experts on Food Security and Nutrition. 2011. *Land tenure and international investments in agriculture.* A report by the High Level Panel of Experts on Food Security and Nutrition of the Committee on World Food Security. Rome: FAO.

Holt-Giménez, E., and M. Altieri. 2012. Agroecology, food sovereignty, and the new Green Revolution. *Agroecology and Sustainable Food Systems* 37:90–102. doi:10.1080/10440046.2012.716388

International Fertilizer Development Center. 2011. *IFDC annual report.* Muscle Shoals: IFDC.

Jayatilaka, R. 1989. The impact of Green Revolution on the Sri Lankan peasantry. *Sri Lanka JSS* 12:21–30.

Jayne, T. S., J. Govereh, M. Wanzala, and M. Demeke. 2003. Fertilizer market development: A comparative analysis of Ethiopia, Kenya, and Zambia. *Food Policy* 28:293–316.

Jiang, D., H. Hengsdijk, T.-B. Dai, W. de Boer, Q.Jing, and W.-X. Cao. 2006. Long-term effects of manure and inorganic fertilizers on yield and soil fertility for a winter wheat-maize system in Jiangsu, China. *Pedosphere* 16(1):25–32. doi:10.1016/s1002-0160(06)60022-2.

Jones-Casey, K., and A. Knox. 2011. *Land administration at a crossroads: Lesson 4.* Ghana: World Resources Institute/ Landesa.

Karbo, N., and W. Agyare. 2002. *Crop livestock systems in Northern Ghana.* Workshop on Crop-Livestock Systems in Dry Savannas of West and Central Africa. Ibadan, Nigeria: IITA.

Kennedy, E., and B. Cogill. 1987. *Income and nutritional effects of the commercialization of agriculture in Southwestern Kenya.* IFPRI Working Paper 63.Washington, DC: IFPRI.

Koopman, J. 2012. Will Africa's Green Revolution squeeze African family farmers to death? Lessons from small-scale high-cost rice production in the Senegal River Valley. *Review of African Political Economy* 39:500–511. doi:10.1080/03056244.2012.711076

Kowal, J. M., and A. H. Kassam. 1978. *Agricultural ecology of Savanna. A study of West Africa*. Oxford, UK: Clarendon Press.

Malley, Z., A. Temu, N. Kinabo, S. Mwigune, and A. Mwageni. 2001. Sowing maize in pits. Farmers' innovation in Southern Tanzania. In *Farmer innovation in Africa: A source of Inspiration* (eds. C. Reij and A. Waters-Bayer), 267–277. London: Earthscan.

Ministry of Food & Agriculture. 2011. Government to spend $57.7 million on 2011 fertilizer subsidy programme. http://mofa.gov.gh/site/?p=2349 (accessed January 1, 2012).

Minot, N., and T. Benson. 2009. *Fertilizer subsidies in Africa*. IFPRI Issue Brief 60. Washington DC: International Food Policy Research Institute.

Morris, M., V. Kelly, R. Kopicki, and D. Byerlee. 2007. *Fertilizer use in African agriculture. Lessons learned and good practice guidelines*. Washington, DC: World Bank.

Moseley, W. G., J. Carney, and L. Becker. 2010. Neoliberal policy, rural livelihoods, and urban food security in West Africa. A comparative study of The Gambia, Cote d'Ivoire, and Mali. *Proceedings of the National Academy of Sciences* 107:5774–5779. doi:10.1073/pnas.0905717107

Neumayer, E. 2003. *Weak versus strong sustainability. Exploring the limits of two opposing paradigms*. Cheltenham, UK: Edward Elgar.

Ngwenya, H., J. Haggman, and J. Ramaru. 2009. Going to scale with facilitation for change: Developing competence to facilitate community emancipation and innovation in South Africa. In *Innovation Africa: Enriching farmers' livelihoods* (eds. N. Sanginga, A. Waters-Bayer, S. Kaaria, J. M. Njuki, and C. Wettasinha), 340–357. London: Earthscan.

Niazi, T. 2004. Rural poverty and the Green Revolution: The lessons from Pakistan. *Journal of Peasant Studies* 31:242–260. doi:10.1080/0306615042000224294

Obeng-Odoom, F. 2013. Underwriting food security the urban way. Lessons from African countries. *Agroecology and Sustainable Food Systems*: 37:614–628. doi:10.1080/21683565.2012.762440

Odame, H., and E. Muange. 2011. Can agro-dealers deliver the Green Revolution in Kenya? *IDS Bulletin* 42(4):78–89. doi:10.1111/j.1759-5436.2011.00238.x

Oduro, A., and I. Osei-Akoto. 2007. *Market participation and rural poverty in Ghana in the era of globalization*. Research Paper No. 2007/70 Helsinki: United Nations University World Institute for Development Economics Research.

Orr, A., B. Mwale, and D. Saiti. 2001. Market liberalisation, household food security and the rural poor in Malawi. *European Journal of Development Research* 13:47–69. doi:10.1080/09578810108426780

Ouedraogo, E., A. Mando, and L. Stroosnijder. 2006. Effects of tillage, organic resources and nitrogen fertilizer on soil carbon dynamics and crop nitrogen uptake in semi-arid West Africa. *Soil and Tillage Research* 91:57–67.

Patel, R. 2012. The long Green Revolution. *Journal of Peasant Studies* 40:1–63. doi:10.1080/03066150.2012.719224

Peters, P. E. 2009. Challenges in land tenure and land reform in Africa. Anthropological contributions. *World Development* 37:1317–1325. doi:10.1016/j.worlddev.2008.08.021

Pinckney, T. C., and P. K. Kimuyu. 1994. Land tenure reform in East Africa: Good, bad or unimportant? *Journal of African Economies* 3:1–28.

Pinitpaitoon, S., A. Suwanarit, and R. W. Bell. 2011. A framework for determining the efficient combination of organic materials and mineral fertilizer applied in maize cropping. *Field Crops Research* 124:302–315. doi:10.1016/j.fcr.2011.06.018

Porter, G., and K. Phillips-Howard. 1997. Comparing contracts. An evaluation of contract farming schemes in Africa. *World Development* 25:227–238.

Pretty, J. 1995. *Regenerating agriculture. Policies and practice for sustainability and self-reliance*. London: Earthscan.

Pretty, J., R. Ruben, and L. A. Thrupp. 2002. Institutional changes and policy reforms. In *Agroecological innovations* (ed. N. Uphoff), 251–260. London: Earthscan.

Prothero, M. 1957. Land use at Soba, Zaria Province, Northern Nigeria. *Economic Geography* 33:72–86.

Rakodi, C. 2002. A capital assets framework for analysing household livelihood strategies. Implications for policy. *Development Policy Review* 17:315–342.

Rasool, R., S. Kukal, and G. Hira. 2008. Soil organic carbon and physical properties as affected by long-term application of FYM and inorganic fertilizers in maize–wheat system. *Soil and Tillage Research* 101:31–36. doi:10.1016/j.still.2008.05.015

Recep, Ç. 2004. Effect of water stress at different development stages on vegetative and reproductive growth of corn. *Field Crops Research* 89:1–16. doi:10.1016/j.fcr.2004.01.005

Reddy, S. 1988. *Use of animal power in West African farming systems. Farm level problems and implications for research: Perspectives from Mali*. Animal power in farming system—Proceedings of West African Animal Traction Networkshop. Freetown: Vieweg.

Reij, C., and A. Waters-Bayer, eds. 2001. *Farmer innovation in Africa: A source of inspiration*. London: Earthscan.

Robertson, B., and P. Pinstrup-Andersen. (2010): Global land acquisition. neo-colonialism or development opportunity? *Food Security* 2:271–283. doi:10.1007/s12571-010-0068-1

Sanchez, P. A. 1994. Tropical soil fertility research: Toward the second paradigm. In *Inaugural and state of the art conferences*. Transactions of the 15th World Congress of Soil Science, 65–88. Acapulco, Mexico.

Sanginga, N., and P. Woomer. 2009. *Integrated soil fertility management in Africa. Principles, practices and developmental process*. Nairobi: Tropical Soil Biology and Fertility Institue of the International Centre for Tropical Agriculture.

Sumberg, J., J. Thompson, and P. Woodhouse. 2012. Why agronomy in the developing world has become contentious. *Agriculture and Human Values* 30:71–83. doi:10.1007/s10460-012-9376-8

Swindell, K. 1985. *Farm labour*. Cambridge, UK: Cambridge University Press.

Tabo, R., A. Bationo, B. Amadou, D. Marchal, F. Lompo, M. Gandah, et al. 2011. Fertilizer microdosing and "warrantage" or inventory credit system to improve food security and farmers' income in West Africa. In *Innovations as key to the Green Revolution in Africa* (eds. A. Bationo, B. Waswa, J. Okeyo, F. Maina, and J. Kihara), 113–121. Amsterdam: Springer. http://dx.doi.org/10.1007/978-90-481-2543-2_10 (accessed June 1, 2013).

Thompson, C. B. 2012. Alliance for a Green Revolution in Africa (AGRA). Advancing the theft of African genetic wealth. *Review of African Political Economy* 39:345–350. doi:10.1080/03056244.2012.688647.

Tittonell, P., M. Corbeels, M. T. van Wijk, B. Vanlauwe, and K. E. Giller.2011. Combining organic and mineral fertilizers for integrated soil fertility management in smallholder farming systems of Kenya: Explorations using the crop-soil model FIELD. *Agronomy Journal* 100:1511–1526

Toenniessen, G., A. Adesina, and J. DeVries. 2008. Building an alliance for a Green Revolution in Africa. *Annals of the New York Academy of Sciences* 1136:233–242. doi:10.1196/annals.1425.028

Uphoff, N. (Ed.) 2002. *Agroecological Innovations: Increasing food production with participatory development*. London: Earthscan

Vanlauwe, B. 2009. ISFM Research at TSBF. The framework, the principles and their application. In *Managing nutrient cycles to sustain soil fertility in Sub-Saharan Africa* (ed. A. Bationo), 25 Mexico: TSBF of CIAT.

Vanlauwe, B., A. Bationo, J. Chianu, K. E. Giller, R. Merckx, U. Mokwunye, et al. 2010. Integrated soil fertility management. Operational definition and consequences for implementation and dissemination. *Outlook on Agriculture* 39(1):17–24.

Vanlauwe, B., J. Chianu, K. E. Giller, R. Merckx, U. Mokwunye, P. Pypers, et al. 2010. *Integrated soil fertility management. Operational definition and consequences for implementation and dissemination*. Working Party on Agricultural Policy and Markets, November 15–17, Washington, DC: IFPRI.

Vanlauwe, B., and S. Zingore, S. 2011. Integrated soil fertility management. An operational definition and consequences for implementation and dissemination. *Better Crops* 95(3):4–7.

Webster, N. 1990. Agrarian relations in Burdwan district, West Bengal: From the economics of green revolution to the politics of Panchayati Raj. *Journal of Contemporary Asia* 20:177–211.

Wiggins, S., G. Argwings-Kodhek, L. Leavy, and C. Poulton. 2011. *Small farm commercialisation in Africa: Reviewing the issues*. FAC working Paper 23. Brighton, UK: Future Agricultures Consortium.

Wiggins, S., and H. Leturque. 2010. *Helping Africa to feed itself: Promoting agriculture to reduce poverty and hunger*. FAC Occasional Paper 002. Brighton, UK: Future Agricultures Consortium.

Wopereis, M., A. Tamélokpo, K. Ezui, D. Gnakpénou, B. Fofana, and H. Breman. 2006. Mineral fertilizer management of maize on farmer fields differing in organic inputs in the West African savanna. *Field Crops Research* 96:355–362. doi:10.1016/j.fcr.2005.08.004

Index

Note: Page numbers in *italics* represent tables
Page numbers in **bold** represent figures
Page numbers followed by 'n' refer to notes